Global Environment
Problems and Policies

Global Environment
Problems and Policies

Edited by

K.R. Gupta
Maria Anna Jankowska
Prasenjit Maiti

ATLANTIC
PUBLISHERS & DISTRIBUTORS (P) LTD

Published by

ATLANTIC

PUBLISHERS & DISTRIBUTORS (P) LTD

B-2, Vishal Enclave, Opp. Rajouri Garden,
New Delhi-110027
Phones : 25413460, 25429987, 25466842

Sales Office
7/22, Ansari Road, Darya Ganj,
New Delhi-110002
Phones : 23273880, 23275880, 23280451
Fax : 91-11-23285873
web : www.atlanticbooks.com
e-mail : info@atlanticbooks.com

ISBN 978-81-269-0845-5 (Vol. I)
ISBN 978-81-269-0849-3 (Set)

Printed in India
at Nice Printing Press, Delhi

Preface

The 20th century has witnessed major developments, which have momentous impact not only on the lives of human beings but also on planet earth—the only known planet in our solar system that supports life by virtue of its unique environmental condition: water, an oxygen-rich atmosphere, and a suitable surface temperature. Limitless greed, reckless consumption of natural resources and unkind treatment meted out to environment has increasingly damaged the earth. These include an unprecedented growth in population and in consumption, rapidly increasing urbanizations, dramatic changes in the global economic system and the revolution in communication technology. We have become increasingly aware that the process of development including industrialization and economic growth has come at a huge cost to society, to the environment and to our very future.

It is ironical that some of the achievements of the 20th century themselves pose the challenges for the present 21st century. As of 2007, the world population has reached 6.68 billion. According to data from the CIA's *World Fact Book* (2007), world human population increase by 211,090 people every day. With 6.68 billion people inhabiting this earth, and the global economy over the next 25 years could expand from $35 trillion in 2005 to $72 trillion in 2030 (World Bank), the pace of human and economic activity is putting pressure on global life support systems.

Today, we witness the depletion of non-renewal natural resources, the continued destruction of life sustaining forests and illegal international trade in protected species of flora and fauna. There is an ongoing depletion of our planet's biological treasures and the threat of acceleration in global warming and in the greenhouse effect. While the monsoon's behaviour is changing sharply, the glaciers are receding because of global warming. The total surface area of glaciers worldwide has decreased 50 per cent since the end of the 19th century. The retreat has been stark in the Himalayas. It may be noted here that the Himalayan glaciers are sensitive indices of climate change due to their typical vulnerability: snow accumulation and melting on these depends on monsoonal precipitation as well as summer temperatures. To quote the Fourth Assessment Report released by IPCC in 2007, "It is likely that future topical cyclones (typhoons and hurricanes) will become more intense, with larger peak wind speeds and more heavy precipitation associated with ongoing increases of sea surface temperatures." Thus, it is likely that the increase in atmospheric temperature resulting in increase in precipitation on glaciers will eventually affect a variety of planning issues, such as the planning and design of hydrological structures, river basin management, flood control and drought management, urban planning and industrial development and above all agricultural policies. To put it simply, climate change is a larger challenge of sustainable development that affect global economic growth.

Besides, there is a rapid globalisation of environmental destruction. The earth is fast losing its treasures. The soil we cultivate, the water we drink and the air we breathe are all polluted. The environmental impact of human activity is no longer restricted to man-made national boundaries.

There is a paramount need to create widespread consciousness of the environment. It must permeate all ages and all sections of society, beginning with the child. The environmental problems call for more than scientific and technological solution. They require the re-establishment of environmental ethics and values practiced in traditional societies, with economic development based on these issues.

For developing awareness among the readers about the preservation of biodiversity and its significance for life, sustainability and equity, and for warning against the consequences of environmental exploitation as well, the present Volumes have been compiled. These have a wide coverage and comprehensively study not only the aforesaid problems but also the policies and schemes adopted relating to the environment.

A major impetus in this regard was provided by the 1987 Report of the World Commission on the Environment and Development (the Brundtland Report) and was accelerated by the Rio Summit in 1992. Although the environmental impact assessment was first formally established in USA in 1969, the new techniques and approaches, policies and agreements have been continuously evolved to prevent damage to existing resources and to promote its conservation and augmentation. In this direction, the latest headway was made in the 12th Conference of Parties to the UN Climate Convention held in Nairobi, Kenya in November 2006, which has decided to set up an adaptation fund to help poor countries adapt to climate change.

The book would be highly useful for students and researchers on Environment. It would be useful to Government executives and NGOs concerned with environment and pollution problems. Since the laymen are deeply interested in clean environment, the book would be of great interest to them.

K.R. Gupta
Maria Anna Jankowska
Prasenjit Maiti

Contents

All articles were originally published in the Electronic Green Journal, ISSN 1076-7975.

Introduction

Global Environmental looks at theoretical and empirical aspects of environmental issues in a way that brings the critical dependencies between the human and physical environment into focus. The social aspects of environmental issues have become ever more prominent since the Second World War, when the impact of weapons of mass destruction and nuclear holocaust made us painfully aware of the politics, economics and sociology that affect the natural environment. Accordingly, the scope of environmental studies has of late become increasingly interdisciplinary, expanding to include exciting new disciplines such as political ecology, environmental economics and social as well as moral ecosophy.

Our unique advantage as Editors of this Volume is that we come from a social science background and can, therefore, utilize strategies from the social sciences to negotiate with the pedagogics of the environmental sciences. We appreciate that even environmental advocacy and activism have their own esoteric aspects that need to be constantly interfaced with the realities and politics of grassroots-level everyday life.

Sustainable environmental strategies have become one of the major concerns of the new millennium at the policy level, even as the global forces of consumerism, resource extraction, mass production and poverty work against sustainability. Such a state of affairs demands that we start exploring viable alternatives to sustain our physical and social environments, as these largely influence critical growth patterns of the global economy. Viable environmentalism then emerges as a serious business that needs to be addressed in terms of workable strategies.

Benefit Monitoring and Evaluation (BME) of environmental development programs is crucial to the physical progress measurement of technical interventions and designs at the level of planning and implementation. It is important to understand that BME studies also include in their purview the issues and concerns of participatory development. Civil society organizations that engage social and environmental entrepreneurs focused on capacity building and awareness-generation more often than not interact with both primary and secondary stakeholders in order to sustain the benefits of strategic environmental initiatives.

For participatory approaches to environmental BME programs to be effective, such endeavors need to be incorporated into the overall project design and undertaken on a regular basis. The results of an initial external evaluation exercise, for instance, can provide strategic input findings that lay the groundwork for follow-up environmental assessments. These lessons in turn can be learned and replicated to refine the key issues and generate new questions during subsequent BME programs. Environmental development projects should simultaneously develop new strategies and tackle problems that significantly affect the stakeholders' perceptions of program-related interventions.

Environmental awareness and education programs must also seek to build and sustain a stock of social capital at the local level that facilitates BME of interventions undertaken during the project cycle. Social capital ensures that the environmental community continues to preserve its collective assets during the phased withdrawal of environmental development programs.

Environmental agencies cannot always effectively address the high expectations of global and local communities without getting involved in empathetic exercises at the strategic level. Environmental monitoring and management efforts need to include local-level realities and be attuned to marginalize local voices in order to succeed at the level of the global. Environmental scientists and engineers should attend community-level interactions to sensitize themselves to the priorities and perceptions of small-scale stakeholders. Gender issues in relation to developmental and environmental priorities must also necessarily inform such choices.

We have tried to include papers in these Volumes that constitute lessons to be replicated elsewhere, in different contexts and under different circumstances. A range of real-world experiences underpinned by scholarly analyses have been collected here from different socio-economic as well as politico-cultural viewpoints, reflecting the state of the environment in the context of pluralist systems and multicultural societies. What comes across clearly in the papers published in these Volumes is the understanding that social and physical environments impact upon one another symbiotically, and therefore must be considered together if effective solutions to environmental problems are to be found. As the Editors we have stressed learning from one another, and this dynamic has guided us through the difficult yet rewarding process of selecting the papers and deciding on the priority areas and academic thrusts of these Volumes. *Global Environment* in totality is a unique initiative: individuals from a mosaic of cultures came together to extend their academic support and intellectual rigor in order to make this venture a success. Intellectual associates gradually turned into intimate comrades during the course of planning and implementing this book project. This is indeed a remarkable achievement in professional and personal terms, and we would be even more delighted if it proves useful in addressing the issues it raises worldwide.

K.R. Gupta
Maria Anna Jankowska
Prasenjit Maiti

Ideology and Politics: Essential Factors in the Path toward Sustainability

1

John Coates, Terry Leahy

With the Kyoto Protocol coming into force we can expect media attention to focus on governments' efforts to reduce global warming. While some have questioned how effective such efforts will be due to the number of countries who have not signed on (notably the USA), the potential effectiveness of efforts to combat climate change, and more generally to achieve a sustainable society, can better be understood in the light of ideological and socio-political contexts. This paper argues that ideology and socio-political structure are essential considerations in the path toward a sustainable society. Four models are discussed in terms of their potential to achieve sustainability.

The review of ecological devastation, much of it occurring in the past 100 years, exposes our economy to be an "extractive economy" (Berry, 1997). An extractive economy depletes non-renewable resources, exploits renewable resources beyond their capacity to survive, and causes irreparable damage to land, sea and air (see Coates, 2003a; Foster, 1999; McLaughlin, 1993; Trainer, 1985). Further, the production of toxins along with industrial and domestic effluent greatly exceeds the healing and regenerating capacities of the Earth (see Colborn, Dumanoski and Myers, 1997). The Earth cannot cope with such excesses as human activity has changed the chemistry of the planet and altered the eco-systems upon which modern civilization depends. In fact, no eco-system on Earth is free from the pervasive influence of chemical discharges (Vitousek, Mooney, Lubchenko, and Melillo, 1997). Accompanying this environmental impoverishment has been human exploitation and impoverishment (see Chossudovsky, 1998; Kassiola, 1990; Korten, 1995; Latouche, 1993). Despite considerable information and public attention to environmental concerns, people at large and many businesses and governments have not been motivated to take these issues seriously and have not engaged in effective action toward sustainable practices.

It is our contention that the major reasons for this lack of concern and action stem from Western society's embeddedness in a particular set of values, beliefs and assumptions, and a socio-political structure, which are at the foundation of public and individual action. It is this embeddedness in the assumptions and beliefs of modernity, which Spretnak refers to as the 'denial of the real' (1997), within the context of liberal capitalism that stands in the way of people and governments developing effective responses to, and also becoming involved in, the promotion of environmental and social justice. This system of beliefs, referred to as modernity, places absolute confidence in technology and science, and has unquestioned confidence in consumer-oriented and market-driven growth and development. Paul Hawken captures this when he states that modernity has "quite naturally produced a dominant commercial culture that believes all resources and social inequities can be resolved through development, invention, high finance and growth—always growth" (1993, p. 5).

Exploitation and destruction, along with development, are the outcomes of modern society and in particular, its values and beliefs, and political structure. While this belief system has had many achievements it has also had its "dark side" (Capra, 1982)—environmental and social injustice. However, many people are so embedded in modernity that they are incapable of recognizing that it is "the structures and processes of everyday life that cause environmental destruction and social injustice" (Coates 2003a, p. 27). Sadly, most people have not explored the assumptions and beliefs which inform their own, and their society's actions. Successful environmental and social justice initiatives will not be forthcoming in the absence of a critical examination of foundational beliefs and socio-political contexts. It is essential to recognize that "Environmental issues are... social and cultural issues" (Rogers, 1994, p. 1). Only when the foundational ideologies and structures are critically examined will society be in a position to adopt an alternative worldview and socio-political structure that are capable of supporting the emergence of a sustainable society.

The Dominant Worldview—Modernity

An exploration of the root causes of the environmental crisis calls into question the values, beliefs and assumptions that inform modern society. Baudrillard (1987, pp. 63-64) views modernity as an ideology which, over recent centuries, has become a "characteristic of civilization" acting as a "regulatory control function". In the absence of an alternative vision the belief in human privilege, progress, and technological innovation contributes to a society complacent about ecological and social exploitation. Catton and Dunlap's (1980, pp. 17-18) analysis of this "Dominant Western Worldview" can be summarized as:

1. People are fundamentally different from, and superior to, all other creatures.
2. People are masters of their destiny, and can use the rest of nature in any way they choose.
3. The world is an endless resource, and thus provides unlimited opportunities.
4. Human ingenuity will solve all problems, and progress need never cease.

This innate bias reflects the beliefs that have informed colonization, industrial and economic development, as well as the response to social problems. Humans (and in some historical contexts Caucasians) thought themselves to be the end and purpose of creation. As a result, the predominant criterion to evaluate not only industrial and agricultural practices, but also the value of plants and animals, was their perceived "use value" for human endeavours. Such anthropocentrism has contributed to a shortsighted, exploitive and unsustainable criterion for progress. Within such an anthropocentric foundation the Earth was seen to be an abundant and unending source of commodities; and progress focussed blindly on converting natural materials (through technology) into consumer goods and very quickly into waste. In fact a healthy economy is seen as "an expanding economy, in which more and more material goods are produced, consumed and discarded" (Clark, 1989, p. 30). Human ingenuity through technology was considered capable of solving all problems that would enable progress to continue unabated.

Emerging within this attitude of superiority and exploitation are very powerful beliefs that govern public, industrial and frequently personal decisions. Economism, progressivism, industrialism, consumerism and individualism serve to set the direction of human actions. Charlene Spretnak's description of modernity includes the following characteristics:

Homo Economicus:	Economic well-being is primary and leads to well-being in other parts of life
Progressivism:	Technology will solve all problems and the human condition will gradually improve through abundance
Industrialism:	Mass production = > abundance = > consumerism = > happiness
Consumerism:	Material goods are the source of happiness
Individualism:	Competition for individual benefit; individual interests take priority over communal interests (1997, pp. 40-41).

The predominance and centrality placed on the role of economics and of economic growth (see Hamilton, 2003) dominates economic and political decision-making. "Economics has become the source of meaning and relation in modern society" (Rogers, 1994, p. 86). Economism is so strong that even some sustainability advocates see the economy as a fixed reality rather than a means to a greater good (for example, see Hart, 2002). Within such a belief system money and its possessors becomes the supreme good (see Marx, 1978), and affluent living standards spurred by incessant and pervasive consumerism overshadow all other concerns. All of these beliefs enable the marketplace to be the major determinant of what happens in a society (Trainer, 2003) and reinforce the belief that abundance (through mass production and mass consumption) will solve all problems.

TV, the dominant medium of Western culture, has become the primary vehicle for socialization (Swimme, 1997). Children in North America watch, on average, more than 3.5 hours of TV each and every day and in the process consume thousands of advertisements. By the time a North American child graduates from high school, she/he has spent more hours absorbing advertisements than listening to their teachers in the classroom. Is it any wonder then that society is focussed on consumerism! For many of today's youth the path to the "good life" can be summarized as obtaining a good education, to secure a good job, so you can earn good money, so you can buy things (Swimme, 1997). Status is conferred upon those in possession of wealth, and money has become a primary determinant of self-worth (Durning, 1991). The illusion created by the dedication to progress and development, and the myth of human betterment through possession of increasing amounts of material creations —more and more stuff prevails.

"We now consume more in one year than we did for the whole period from the birth of Christ to the dawn of the industrial revolution" (Peter Russell 1998, p. 43).

At the very core of this worldview, and supporting these beliefs, are the foundational assumptions of modernity—dualism, domination and determinism (see Coates, 2003a). Within this worldview humans are separate and distinct from each other, from nature, and from 'God', and domination (people over nature; rational over emotional, and people over people) over a mechanistic universe is seen as natural. While media attention has exposed the exploitation of nature, the bias toward dualism, unending progress, economic primacy, and consumerism have evaded mainstream criticism.

Dualism:	Nature remains in background; a resource for human need

	People and their lives are divided into distinct parts
	People are separate and distinct (from nature and each other)
Domination:	Values remain primarily human-centered
	Socialized passivity; powerlessness in relation to BIG business
Determinism:	Universe (Earth) as biological machine; as a collection of objects
	Universe is fixed, never changing
	Outcome/performance-oriented approaches to intervention
	Earth, society and people are aggregates (collection of objects)
	Scientism, objectivism, managerialism, efficiency.

Modern professions developed within society to serve the needs of progress and industrial development and, as a result, share the same foundational beliefs. The growth-oriented, acquisitive, human-centered, dualistic bias of modernism remains unquestioned. Most professions have not been conceptualized independent of modernity and tend to 'psychologize' problems rather than question foundational assumptions. Many people have been "seduced into docility and willing compliance" (Trainer, 2003) by "culturally conditioned perceptions" (Livingston, 1994), the absence of a viable alternative or focussed resistance.

Many educational institutions, spurred on by the globalization of trade, are preoccupied with preparing workers and consumers, rather than a deep exploration of critical social issues (Trainer, 2003). Schools emphasize order, competition, hard work, individualism, and the belief that a school grade or diploma entitles one to social privilege. An ecologically just and socially just society is not possible when social life is dominated by market forces, profit, economic growth, and increasing standards of luxury; unrestrained consumerism leads to unrestrained exploitation. A satisfactory society, one that concerns itself with well-being, must have a concern for the public good, for the well-being of all. A sustainable society can be seen to enable imaginative and progressive change (Hill, 2004), and to enhance the well-being of individuals, society, and the Earth. Public standards and public institutions can maintain a sense of social solidarity in which we are willing to contribute to the well-being of all.

An Alternative Worldview

To secure an effective solution to environmental and social injustices an alternative foundation of beliefs and values is essential. Such a perspective has been developed by Berry (1988) and is reflected in the work of others including Coates (2003a), Devall and Sessions (1985), Plumwood (1993), Roszak (1992), and Suzuki and Knudtson (1993). Such a view considers every person and all of nature to be important participants in the life of the planet; humanity emerged within nature and human development, perhaps even our survival, is dependent upon a thriving Earth community. Replacing the dualism, domination and determinism of modernity are an alternative set of assumptions which form the foundation of an alternative worldview, that of an emerging or unfolding universe. Core assumptions include:

Interdependence:	Whole system consciousness
	Connected and overlapping subsystems
Care:	Earth exists to be shared by all things
	Diversity and complexity exist within and among systems
	Limits exist so other species may thrive
Emergence:	The Universe, Earth, and humanity are unfolding
	Every being plays a role in their own evolution and that of the whole.

Within such a perspective, Earth is considered as sacred, as a nurturer of higher existence, as the context within which life has emerged; it is not simply a resource. People are embedded within nature, and connected to the interdependent life of the planet. Earth is a geological and biological system. Within this overarching whole there are many overlapping and interlocking subsystems (ecologies, catchments, villages, economies, etc.). While all objects and living beings may be conceived as independent in particular contexts, they are also tied into many systems. Given the power of human technology, our survival and that of other species requires that we pay careful attention to these interactions (including the known, unknown and unimagined). Changes made to any element can have cascading and potentially disastrous effects.

This moves to the ethic of care both out of enlightened self-interest and the right of other species to continue to exist (we do wonder though, about such things as smallpox and AIDS/HIV). While good science can provide a critical analysis of the connections that may or may not exist, our chances of survival as a species depend on humans developing our capacities to care for, and feel empathy for, other people and other living beings. Together we share many of the systems that make up life on Earth, and our care and attention to their needs can help us to pay attention in ways that will end up helping us to survive. Further, it is becoming increasingly clear that caring for other people and other beings can be a great source of pleasure and fulfillment. Caring for 'others' can be an expression of our spirituality as it helps us to transcend an exclusive self-interest. A basic expression of love for other beings is to share the planet and its resources, recognizing that other species have a right to live—it is a co-evolutionary path (Roszak, 1992; Plumwood, 1993).

What emerges from interactions with other systems and from the operation of the system itself are always developments that cannot be deducted from an examination of the elements of a system. These emerging properties are ongoing; the Universe, Earth and humanity are unfolding. Every being plays a role in their own emergence and that of the whole.

These assumptions demand a conscious participation in the personal and communal struggle to live in the knowledge of our essential connectedness to the Earth. We have a moral responsibility to live harmoniously within the biosphere (Berger and Kelly, 1993). Such a communal worldview sees individuality in the context of the whole, and can serve as motivation to bring about a shift in personal lifestyle and social organization that is both sustainable and socially just. This represents a significant shift in consciousness, a refreshing counter to modernity's human-centered exploitation. However, motivation is not enough (see Revington, n.d.), we must explore the nature of socio-political structures in which such alternative values and beliefs can be put into practice. A review of socio-political structures can help to determine in which one, or ones, sustainability and social justice can be realized.

Four Models and Sustainability

Broadly speaking, four models of social-political structures have been proposed, whether by environmentalists or by branches of the left. Each of these can be regarded as an ideal type so proposals may be developed that draw on one or more models.

Model One: Regulating the Capitalist Economy

In this model, environmental sustainability is achieved by regulating the mechanisms of the capitalist marketplace. One method is interdiction—rules, laws, administrative principles that ban certain kinds of conduct (such as release of pollutants). Another is adjustment to incentives—systems of taxation, subsidies and tariffs. Amory Lovins is widely known for this position (Hawken, Lovins, and Lovins, 2000); many others propose some version (Andersen, 1994; Beck, 1995a; 1995b; Goldsmith, 1988; Pearce, Markandya, and Barbier, 1989; Sachs, Loske and Linz, 1998). Current efforts to meet the expectations of the Kyoto Protocol are most likely to fit this model.

The structure of capitalism as an economy based on competitive private ownership is a problem for this model (McLaughlin, 1993). Managers of any business are required to maximize profitability. They can do this by various means including expanding markets, cutting costs, bringing in new technology, and saving on labor costs. In a competitive environment, any failure to maximize profitability can cause business collapse. Accordingly, every manager tries to *externalize* environmental costs, doing their utmost to avoid environmental regulation. Adding to this, shareholders and managers of private firms are the wealthiest people, and the ones who control media and who can most readily afford political campaigns to pressure governments for non-binding and minimal regulations in efforts to retain profits.

Second, capitalism also creates galloping and inevitable growth—firms compete to reduce labor costs by bringing in more productive technology and increasing markets. The result is constant expansion and the threat of growing unemployment if growth slows.

A third problem is the interests of consumers. When Marxists say that labor is "alienated" they mean that people do not take creative pleasure in their paid work; those who own the company or control their government department dictate what they do at work. In capitalism, to have access to goods and services, people must sell their labor power (Marx, 1978); hence the cultural dominance of consumerism. Material consumption is an arena of choice and power that is valued precisely because of the absence of choice, creativity and power in the world of paid work. In turn the capitalist economy has come to depend on the expansion of the market that comes from pressure for wages (Cardan, 1974).

Any serious regulation of the capitalist economy would result in increasing restriction of leisure and consumer choice with lower real wages and fewer consumer goods. This would be experienced as an unwelcome strangling of freedom of expression. In the regulated capitalist economy, people would still have little control of their work and consumers would have lots of reasons to resist environmental policies. As today, this experience would provide a powerful motive for the accumulation of private consumer goods—a compensation for paid work.

Summarizing these objections to the first model; regulated capitalism leaves intact the key structures of ownership and paid work that are at the heart of capitalism. Yet it is these very structures that are the wellsprings of the resistance to environmental regulation and explain the political force of this resistance. Any serious attempt to implement this model could either founder on this resistance—with a return to business as usual—or be forced to move on to deeper attacks on the structures of capitalism, thereby changing into one of the other models we will consider.

Model Two: The Mixed Economy Model

The mixed economy model implies radical change from current socio-political arrangements. In this model a democratic government supervises three arenas. It regulates the private sector, which produces consumer goods and provides inputs to public works. Second, government handles environmental repair and public infrastructure. Third, government funds (through Guaranteed Adequate Income—GAI),[1] but does not control, a voluntary or "autonomous" sector, which engages in such things as community projects and childcare.

This attractive model has been very popular both with environmentalists (see Goodin, 1992; Leahy, 2001; Porritt, 1990; Tokar, 1987; Trainer, 1995) and before that with the new left (Gorz, 1982). The government sector would expand, as many firms would not survive strong environmental regulation. Unemployment, however, would not be a problem as the government finds people a job, or funds them on a guaranteed adequate income or shortens the working week.

Some early criticisms of this model still seem apt. Frankel (1983, 1987) maintains that capitalist companies cannot easily cope with the higher taxes and greater controls that this model proposes. For example, even in today's capitalist world companies encounter problems of international competition and overproduction. The most likely result would be bankruptcy for many companies, with capital flight likely. While the government could employ displaced workers, it is probable that taxation of a struggling private sector could not provide sufficient funds for this.

Sustainability requires zero or negative growth. Given increasing productivity, the only way to achieve zero growth would be to continually reduce employment in the private economy. However, this private economy would be the source of consumer goods and inputs to the other sectors of the economy. So the number of people fully employed in the private economy falls at the same time as the distribution of its products, to people who are not part of its paid workforce, must continually increase. This situation would be very difficult to handle politically.

Another problem with the mixed economy model is the role of the voluntary sector. While people receiving the guaranteed adequate income could be expected to volunteer for socially useful work, they are not supervised or made to work. The problem can be viewed as a dilemma.

A. The guaranteed adequate income enables a comfortable lifestyle and people are provided with tools and resources to help them make a useful contribution.

- The GAI draws people out of the consumerist lifestyle and soaks up unemployment.
- Yet many leave paid employment to escape hierarchical supervision.
- The experience of democratic control in GAI community groups undermines discipline in the paid workforce.

OR

B. The guaranteed income is less than the lowest wage and does not allow a comfortable lifestyle.

- Those forced onto the GAI through contractions of the private economy remain disaffected.
- The GAI is experienced as low status.
- There is insufficient funding for the voluntary sector to make a meaningful contribution.

While strong commitment may lead to progress at resolving these concerns, a central problem remains—at least two-thirds of people's experience of work in this model is still an experience of taking orders, whether within private firms or government bureaucracies. Because consumer spending is always the legitimate reward for alienated work, pressure to expand spending and growth would continually plague the model in practice.

Model Three: The Nationalization Plus Democracy Model

In the nationalization plus democracy model, democratic control guides a soviet style economy (Commoner, 1990; Martell, 1994; Pepper, 1993; Resistance, 1999; Weston, 1986). Government owns most businesses, but does not control them completely as workers also participate through democratic control of the workplace. Whereas representative democracy was quite nominal in the Soviet Union, current proponents of this model envisage representative democracy as substantial and central to this model. The community at large democratically directs major production decisions and environmental planning.

In this model there would be no problem with businesses going broke and laying off employees. While some industries would be abandoned, others would be taken over by government ownership. New government enterprises would create environmentalist infrastructure. Taxes would not come from private businesses—as in the mixed economy model; instead the government would fund central planning decisions. The problems of the guaranteed adequate income would vanish, as there would be employment for community tasks, not voluntary work on a guaranteed income.

Most people today believe a nationalized economy would curtail political freedom. Every productive organization that could disseminate ideas would be owned by the state.

A more central problem is whether it is politically viable to mix local democratic control of production with national control of the economy. As Cardan (1974), Castoriadis (1987), and Hardt and Negri (2000) have pointed out, capitalism faces a problem like this continuously. Modern production requires workers to make decisions and participate responsibly. Yet these elements of workers' control have to be continually squashed lest they undermine management. The nationalization plus democracy model replicates this problem, presenting its own dilemma.

- Local democracy really means something. Workers use their discretion to subvert and sabotage government directives—getting extra income through corruption or making their own decisions about what is useful production.

OR

- Government control is sufficiently far reaching to prevent any sabotage or subversion. It also prevents any significant workers' control. There is the same disaffection that plagued the Soviet economy.

Another issue is that the economic problems of the Soviet Union would likely be replicated. Feher, Heller and Marcus discuss the problems of Eastern Europe in the Soviet era (1983). At the local level, each management unit tries to increase their share of national resources, in de facto competition with other units to make sure they can carry out the tasks government has set them. They inflate the importance of their projects and understate the costs in an effort to get the government committed so it will make sense to put more into the project at a later date. The bureaucrats respond by discounting all claims. But by how much? The end result is massive inefficiency. Systematic hoarding is the norm—government may not be able to supply resources when needed. Top planning authorities would favor projects that were guaranteed to work out as predicted, frequently at the expense of local democracy. Producer goods and public infrastructure would receive most funding. Further, the quality and durability of all goods and services would suffer without marketplace competition. None of this would help to reduce waste.

In this model, control over labor through the wage would be the key to implementation of planning decisions. People would be working for a wage to get access to goods and services produced by other units. Wages, private income, or income from corruption would still be the main compensation for the submission that goes with paid labor. Each worker would aim to consume more while the government would want to cut back to attain environmental goals. More responsible managers would be rewarded with higher wages, a message that the consumer lifestyle is a mark of social status, power and community trust.

Model Four: The Gift Economy Model

In a gift economy people do not work to get money to buy things. What people get is what they make themselves or what is given to them. People would volunteer their work because they realized a job needed to be done or would be interesting. Social status would reward those who provide products and services. In the gift economy production is not owned by private shareholders or government but by a patchwork of clubs, societies and federated hobby groups. The gift economy model is one kind of anarchist stateless utopia (Guerin, 1970), the term pioneered by the Situationists (Debord, 1977; Plant, 1992; Vaneigem, 1983; more recently Bey, 1991, 1994; P.M., 1985; and Wilson, 1998). Environmentalist anarchists close to this are Bookchin (1971) and Purchase (1994). Other anarchist environmentalists favor self-sufficient communes (Bahro, 1986; Allaby and Bunyard, 1980) or a return to a hunting and gathering existence (Zerzan, 1994). These latter perspectives are usually addressed in critiques of anarchism (Dobson, 1990; Pepper, 1993; Plumwood, 2002).

The gift economy model would occur if the mixed economy or the nationalized economy was unable to contain its contradictions, and moved to new ground rather than back to capitalism. A gift economy is seen to be more compatible with ecological imperatives than capitalism or nationalization for several reasons.

- The producers themselves—to save effort, would reduce useless production. It would make no sense to work hard making useless items that no one else needed and that you did not enjoy making.
- With complete control at the point of production, workers in a gift economy would be able to take steps

to avoid environmental harm to themselves and their communities.

- Producers in a gift economy would produce environmentally sensible items and services to maximize the social value of their gifts; there would be no acclaim if they damaged the environments of their own and other communities.

The gift economy would make it easier to transform technology and infrastructure for sustainability and to accept the sacrifices that would go with this as work is the key arena for creativity and participation. Fulfillment through consumption is less relevant, and creating an environmentally sustainable infrastructure would be an exciting task. If producers organized and controlled the distribution of their products, the need for state intervention, and the size of government could be significantly reduced.

A state depends on employees who will obey central directives to implement its decisions. In a gift economy all people would be supplied by the gifts of a multiplicity of independent producers, and each of these producer groups would be making their own decisions about distribution.

For such a system to work to produce roughly equal outcomes there would have to be a cultural commitment to equality. Equality could not arise from a rationed out impartiality—there would be no central authority to determine equal shares. Instead it would come about through relations with *particular* others that in aggregate were not balanced to the disadvantage of any gender, orientation or cultural/ethnic group. So people would have to be motivated by a generous and sympathetic benevolence. Empathy and aesthetic appreciation would also guide our responses to other species.

Of course such a model is not without its critics—the impossibility of the gift economy is the stock in trade of ideology.

A common objection is that coordination requires authoritative and coercive centralized organization (Pepper, 1993). The gift economy achieves coordination through two factors. First, independent and multi-pronged collectives of media, research and administrative workers keep other producers' collectives up to date with what is required and by whom. Second, the aim of producers' collectives is to ensure an equitable outcome for society as a whole.

Another critique of stateless utopias is that they imply a socially divided populace (Frankel, 1983) where an insular and smug parochialism would prevent an evenhanded distribution. However, the gift economy is not a communalist utopia; it is not "bioregionalism" (Sale, 1991). In the gift economy networks of geographically overlapping producers' collectives would work to sort out problems and secure a fair share.

It is argued that only a democratic state can ensure the rights and well-being of minorities. However, in today's liberal democracies the majority supports these rights. A gift economy would do no worse. In a gift economy the same majority would be implementing these same policies: through the defense of minorities from violence by voluntary collectives of peacekeepers, and through community processes of justice and social work.

Critiques of the gift economy in terms of human nature are common, and many people assume that a competitive human nature would destroy any society that had to depend on egalitarian generosity to work. We can accept that human nature *as it is now socially constructed* operates to prevent a gift economy. For a gift economy to work a cultural shift must take place.

Getting to the Gift Economy

The gift economy, at the present time, does not exist in embryo within capitalism as pure enclaves of a new mode of production. The social transformation and shift in worldview required to move a capitalist society toward one based on a gift economy is substantial (to put it mildly). However, we see the gift economy as the social structure best able to support sustainability. To many people in a capitalist society the gift economy may appear unnatural and unrealistic or, at best, naïve. However, capitalist culture, along with the economic requirements of effective operation within a capitalist economy, produces a variety of hybrid situations. While aspects of the capitalist authoritarian mode of production dominate, some aspects of a gift economy operate to further the goals of the left social movements (Leahy, 2004; Mollison, 1988; Trainer, 1995). Many movements in our society reflect elements of the gift economy already—coops, credit unions, Local Exchange Trading Systems (LETS), Fair Trade Coops, Living Simply (Elgin, 1993), Permaculture, and community gardens, to name a few. The proliferation of these 'alternative' initiatives reflects the efforts of thousands, if not millions, of people to promote the values of community, interdependence, sustainability, and equality. These initiatives promote individuals joining together to create meaning and economic well-being with less and less reliance on mainstream economic structures. In this way they can be seen as precursors of the gift economy.

We suggest that the arrival of the gift economy will occur through the development of these alternatives, cooperative and meaning filled activities. While most will be unique and attend to local needs and resources, some will become very successful, catch on and be imitated by others. Swimme (1997) refers to these as "key attractors". Some of these successful structures will become so substantial that they will form an infrastructure around which new communities and new social structures will form. Such a transformation will proceed more rapidly if the existing socio-economic order breaks down, or if attempts to develop mixed economy or nationalization models fail; if the extractive and exploitive nature of our capitalist economy prevails, one might be very confident of this eventuality.

In the interim we can encourage the development of community-based initiatives that provide meaning and control, that enable local resources to be used by local residents to create local benefit where local benefit includes the needs of all species and future generations. Since many people raised in a capitalist and individualist society will be skeptical, it is essential that alternatives demonstrate success and improvements to the quality of life. In the absence of demonstrated effectiveness, it is unlikely that the necessary critical mass will be secured. A number of actions and practices can foster the knowledge and values that are consistent with the emergence of a gift economy. These include:

- Bringing to public attention the unreasonableness and severe destruction of the present organization and functioning of society can serve to open people to consider alternative world views and ideals that improve the well-being of all, and see human actions and social structures, along with the 'rest of nature', as part of Earth's evolutionary unfolding. The development of an ecological or global consciousness emerges from a critique of human superiority and exclusively human-centered valuations, and an awareness of the injustice and exploitation inherent in modern society.
- Link environmental and social justice issues, environmental decline is almost always associated with social injustice (Coates, 2003b). The reality of those who face severe distress, exclusion and depression is so often a product of relentless low standards of living and societal marginalization.
- Create opportunities for meaningful activity outside of consumerism and materialism. It is essential to counter the problem of the alienated worker where consuming and shopping are the primary, if not the only, areas of choice that many people have. Opportunities can be developed so that commitments and a sense of fulfillment can be obtained from volunteering, socially responsible actions, and participation in community projects.
- Promote the development of communities and alternative organizations that are inclusive, egalitarian and supportive of the creative potential in each person, and supportive of ecological ways of life. These can be found in such movements as Voluntary Simplicity (Elgin, 1993), Engaged Living (Robin, 1999), Cooperacy (Hunter, Bailey & Taylor, 1997), The Simpler Way (Trainer, n.d.), Eco-Villages, Permaculture, and cooperatives.

- Opt out of corporatist processes by joining a co-op, boycotting, and challenging retailers to sell local and Fair Trade products. Small meaningful steps that release the potential of every being are to be celebrated as they demonstrate that fulfillment comes from acting on life's purpose and living in harmony with the web of all life. Complexity science reminds us not to underestimate the potential impact of one person's small steps taken in community with others.
- Changes in socialization methods and family dynamics can nurture more generous, confident and caring adults. Research has shown that indulging infant can nurture generosity and confidence and early childhood needs (Hamilton, 1981). Traditional child rearing in the West has produced adults who are continuously anxious, and who amass and hoard to feel secure. Also as fathers become more involved in the care of young children and develop close nurturing relationship with their children, boys in particular, are more likely to take a more caring role in looking after the community and the planet (Chodorow, 1974). Further, an experience of family and community in which men and women are equal partners is necessary if people are to grow up without expecting someone to be the boss. Firestone (1972) recognized this connection between authority in the family and authoritarian structures in society. To institute the gift economy we must implement the feminist demand for gender parity.

Several of these keys to the gift economy are being implemented now as part of a cultural transition.

Conclusion

As media attention to environmental issues grows, public concern is awakened to find solutions to problems such as species extinction, climate change, and pollution. This paper examined the dominant ideology of modernity and four major socio-political models, in regards to their potential to bring about a sustainable society. A shift to sustainability is unlikely to be achieved without a major shift in ideology or consciousness, away from modernity's dualism, superiority and human-centered exploitation, toward a foundation based on interdependence, care and emergence. Such a radical shift in ideology needs to be accompanied by a consideration of socio-political structures that can be more supportive of sustainability.

A review of four models—Regulated Capitalism, Mixed-Economy, Nationalization plus Democracy, and Gift Economy, leads us to conclude that the Gift Economy is more compatible with ecological imperatives. However, unlikely such a model may appear to members of 'modern' capitalist society, precursors of the gift economy currently exist and it can emerge as a viable model, if not the only one without built in contradictions to sustainability. The limitations of socio-political models must be considered if effective environmental interventions aimed at reducing ecological degradation, and promoting sustainability and ecological well-being, are to be achieved. Hopefully this article can serve to initiate further dialogue on the importance of ideology and politics in the search for a viable path toward sustainability.

Endnote

1. GAI as used in this paper represents Guaranteed Adequate Income. Although there are similarities the use of GAI in this paper should not be confused with Guaranteed Annual Income that may or may not be adequate (above the poverty line).

References

Allaby, M. and Bunyard, P. (1980). *The Politics of Self-sufficiency.* Toronto: Oxford University Press.

Andersen, M.S. (1994). *Governance by Green Taxes.* Manchester: Manchester University Press.

Bahro, R. (1986). *Building the Green Movement.* London: Heretic Books.

Baudrillard, J. (1987). Modernity. *Canadian Journal of Political and Social Theory Revue Canadienne de Theorie Politique et Sociale* XI(3): 63-72.

Beck, U. (1995a). *Ecological Politics in an Age of Risk.* Cambridge: Polity Press.

Beck, U. (1995b). *Ecological Enlightenment: Essays on the Politics of the Risk Society.* New Jersey: Humanities Press.

Berger, R. and Kelly, J. (1993). "Habitat Destruction Syndrome", *Social Work*, 38(5): 521-526.

Berry, T. (1988). *The Dream of the Earth*. San Francisco: Sierra Club.

Berry, T. (1997). "Annual Colloquium with Thomas Berry", Holy Cross Centre for Ecology and Spirituality. Port Burwell, Ontario, June 13-15.

Bey, H. (1991). *T.A.Z. The Temporary Autonomous Zone, Ontological Anarchy, Poetic Terrorism*. New York: Autonomedia.

Bey, H. (1994). *Immediatism*. Edinburgh: AK Press.

Bookchin, M. (1971). *Post-scarcity Anarchism*. San Francisco: Ramparts Press.

Capra, F. (1982). *The Turning Point*. New York: Simon and Schuster.

Cardan, P. (1974). *Modern Capitalism and Revolution*. London: Solidarity.

Castoriadis, C. (1987). *The Imaginary Institution of Society*. K. Blamey (Trans.). London: Polity Press.

Catton, William Jr. and Dunlap, Riley E. (1980). "A New Ecological Paradigm for Post-exuberant Paradigm", *American Behavioral Scientist*, 24(1): 15-47

Chodorow, N. (1974). "Family Structure and Feminine Personality", in M.Z. Rosaldo and L. Lamphere (eds.), *Woman, Culture and Society*. Chicago: Stanford University Press.

Chossudovsky, M. (1998). *The Globalisation of Poverty*. Halifax: Fernwood.

Clark, M. (1989). *Ariadne's Thread: The Search for New Modes of Thinking*. New York: St. Martin's Press.

Coates, J. (2003a). *Ecology and Social Work: Toward a New Paradigm*. Halifax: Fernwood Press.

Coates, J. (2003b). "Exploring the Roots of the Environmental Crisis: Opportunity for Social Transformation", *Critical Social Work*, 3(1): 44-66.

Colborn, T., Dumanoski, D. and Myers, J. (1997). *Our Stolen Future*. New York: Penguin.

Commoner, B. (1990). *Making Peace with the Planet*. New York: Pantheon.

Debord, G. (1977). *Society of the Spectacle*. Detroit: Black & Red.

Devall, B. and Sessions, G. (1985). *Deep Ecology: Living as if Nature Mattered*. Salt Lake City: Peregrine Smith Books.

Dobson, A. (1990). *Green Political Thought*. London: Unwin Hyman.

Durning, A. (1991). "Asking how much is Enough", in Lester Brown et al, (eds.) *State of the World 1991*. New York: Norton. pp. 153-169.

Elgin, D. (1993). *Voluntary Simplicity*. New York: William Morrow and Co.

Feher, F., Heller, A. and Markus, G. (1983). *Dictatorship Over Needs: An Analysis of Soviet Societies*. Oxford: Basil Blackwell.

Firestone, S. (1972). *The Dialectic of Sex: The Case for Feminist Revolution*. New York: Paladin.

Foster, J. (1999). *The Vulnerable Planet: A Short History of the Environment*. New York: Monthly Review Press.

Frankel, B. (1987). *The Post-industrial Utopians*. Cambridge: Polity Press.

Frankel, B. (1983). *Beyond the State? Dominant Theories and Socialist Strategies*. London: Macmillan.

Goldsmith, E. (1988). *The Great U-turn: De-industrializing Society*. Hartland, Bideford, Devon: Green Books.

Goodin, R.E. (1992). *Green Political Theory*. London: Polity Press.

Gorz, A. (1982). *Farewell to the Working Class: An Essay on Post-Industrial Socialism,* M. Sonenscher (Trans.). London: Pluto.

Guerin, D. (1970). *Anarchism* (M. Klopper, Trans). New York: Monthly Review Press.

Hamilton, A. (1981). *Nature and Nurture: Aboriginal Child-rearing in North-central Arnhem Land*. Canberra: Australian Institute of Aboriginal Studies.

Hamilton, C. (2003). *Growth Fetish*. Crows Nest, NSW: Allen and Unwin.

Hardt, M. and Negri, A. (2000). *Empire*. Cambridge: Harvard University Press.

Hart, M. (2002). *Seeking Mino-Pimatisiwin: An Aboriginal Approach to Helping*. Halifax: Fernwood Press.

Hawken, P. (1993). *The Ecology of Commerce*. New York: Harper Business.

Hawken, P., Lovins, A.B. and Lovins, H. (2000). *Natural Capitalism: The Next Industrial Revolution*. London: Earthscan.

Hill, S. (2004). "Social Ecology as a Framework for Understanding and Working with Social Capital and Sustainability", in A. Dale & J. Onyx (eds.). *A Dynamic Balance: Social Capital and Sustainability within Rural Communities*. Vancouver, BC: UBC Press.

Hunter, D., Bailey, A. and Taylor, B. (1997). *Cooperacy: A New Way of Being at Work*. Tandem Press: Auckland.

Kassiola, J. (1990). *The Death of Industrial Civilization: The Limits to Economic Growth and the Repoliticization of Advanced Industrial Societies*. Albany: SUNY Press.

Korten, D. (1995). *When Corporations Rule the World.* San Francisco: Kumarian Press and Berrett-Koehler Publishers.

Latouche, S. (1993). *In the Wake of the Affluent Society: An Exploration of Post-development* (M. O'Connor and R. Arnoux, Trans.). London: Zed Books. (Original work published in 1991)

Leahy, T. (2001). *On the Edge of Utopia: A Letter to the Green Parties.* Retrieved March 10, 2006, from www.octapod.org/gifteconomy

Leahy, T. (2004). *Anarchist and Hybrid Strategies.* Retrieved March 10, 2006, from www.octapod.org/gifteconomy

Livingston, J. (1994). *Rogue primate: An Exploration of Human Domestication.* Toronto: Key Porter Books.

Martell, L. (1994). *Ecology and Society: An Introduction.* Amherst: University of Massachusetts Press.

Marx, K. (1978). *The Marx-Engels Reader.* R.C. Tucker (ed.), New York: W.W. Norton.

McLaughlin, A. (1993). *Regarding Nature: Industrialism and Deep Ecology.* Albany, NY: State University of New York Press.

Mollison, B. (1988). *Permaculture: A Designer's Manual.* Tyalgum, Australia: Tagari Publications.

P.M. (1985). *Bolo'bolo.* New York: Semiotext(e).

Pearce, D., Markandya, A. and Barbier, E. (1989). *Blueprint for a Green Economy.* London: Earthscan.

Pepper, D. (1993). *Eco-socialism: From Deep Ecology to Social Justice.* New York: Routledge.

Plant, S. (1992). *The Most Radical Gesture.* London: Routledge.

Plumwood, V. (1993). *Feminism and the Mastery of Nature.* New York: Routledge.

Plumwood, V. (2002). *Environmental Culture: The Ecological Crisis of Reason.* London: Routledge.

Porritt, J. (1990). *Seeing Green: The Politics of Ecology Explained.* Oxford: Blackwell.

Purchase, G. (1994). *Anarchism and Environmental Survival.* Tucson, Arizona: See Sharp Press.

Resistance Books (1999). *Environment, Capitalism and Socialism.* Sydney: Resistance Books.

Revington, J. (n.d.) *Deep Ecology is not Enough.* Retrieved April 6, 2006, from http://www.rainforestinfo.org.au/deep-eco/deep.htm

Robin, V. (1999). *Engaged Living.* Retrieved April 6, 2006, from http://www.newroadmap.org/vrengaged.asp

Rogers, R. (1994). *Nature and the Crisis of Modernity*. Montreal: Black Rose.

Roszak, T. (1992). *The Voice of the Earth*. Toronto: Simon and Schuster.

Russell, P. (1998). *Waking Up in Time: Finding Inner Peace in Times of Accelerating Change*. Novato, CA: Origin Press.

Sachs, W., Loske, R. and Linz, M. (1998). *Greening the North: A Post-industrial Blueprint for Ecology and Equity*. Londcn: Zed Books.

Sale, K. (1991). *Dwellers in the Land: Bioregional Vision*. Philadephia: New Society.

Spretnak, C. (1997). *The Resurgence of the Real*. Addison-Wesley: Don Mills, Ontario.

Suzuki, D. and Knudston, D. (1993). *Wisdom of the Elders: Honouring Sacred Native Visions of Nature*. Toronto: Bantam Books.

Swimme, B. (1997). "Where does Your Faith Fit in the Cosmos?", *U.S. Catholic*, June, 8-16.

Tokar, B. (1987). *The Green Alternative: Creating an Ecological Future*. San Pedro, Calif.: R&E Miles.

Trainer, F.E. (T.). (1985). *Abandon Affluence*. London: Zed Books.

Trainer, F.E. (T.). (n.d.*)* *The Simpler Way*. Retrieved April 6, 2006, from www.arts.unsw.edu.au/tsw/

Trainer, T. (F.E.). (1995). *The Conserver Society: Alternatives for Sustainability*. London: Zed Books.

Trainer, F.E. (T.). (2003). *The Simpler Way: Our Global Situation*. Retrieved April 6, 2006, from http://futurepositive.synearth.net/2003/01/20.

Vaneigem, R. (1983). *The Revolution of Everyday Life*. London: Left Bank Books and Rebel Press.

Vitousek, P., Mooney, H.A., Lubchenko, J. and Melillo, J. (1997). "Human Domination of Earth's Ecosystems", *Science* 277(July) 494-499.

Weston, J. (ed.). (1986). *Red and Green*. London: Pluto.

Wilson, P.L. (1998). *Escape from the Nineteenth Century and Other Essays*. New York: Autonomedia.

Zerzan, J. (1994). *Future Primitive*. New York: Autonomedia and Anarchy, C.A.L. Press.

Where the Green Is: Examining the Paradox of Environmentally Conscious Consumption

2

Annie Muldoon

There is much debate about the value of "green consumerism". Critics claim that environmentally conscientious shopping has negligible effects, does not address wider issues relating to the creation of needs and capitalism, and has been co-opted by advertisers as a marketing technique. Proponents argue that it provides a forum where consumers can actualize their beliefs about the need for environmental awareness and protection through their purchasing choices. The focus of this paper is to outline these positions while examining the merits and shortcomings of green consumerism. Although this phenomenon does not necessarily question the "assumption of consumption", it does provide a space for environmental activism for individuals who may not wish to participate in deeper ecological activities. Green consumerism would be aided by government intervention in the form of a regulatory body that would guarantee that "green" products have met strict environmental standards.

Introduction

This is not the article I thought I would be writing. I had planned to spend these pages exploring the environmental possibilities contained within the individual consumption of goods. I thought I would begin this paper with a couple of pages about ecological theory, then move into a tight analysis about the opportunity for consumers to make their collective voices heard through "green" product purchases. 'Green consumerism' is defined as "the purchasing and non-purchasing decisions made by consumers, based at least partly on environmental or social criteria" (Peattie, 1992, p. 118). I thought I would spend the majority of the paper listing those companies with questionable environmental records, and illustrate how consumer dollars are being used to pollute rivers, create toxic waste and destroy natural spaces. I pictured making statements like 'did you realize that Company A directly contributes to the destruction of virgin rain forests?' or that 'Company B throws out more waste that New York City?' Finally, to offer my readers hope, I would name companies that not only possessed unblemished environmental histories, but that also offered environmentally-friendly products at reasonable prices. I would provide a condensed version of the paper to all my colleagues and friends, who could coast into their summers confident that their meager budgets were being used to encourage corporate environmental responsibility.

Writing this paper has gone a little differently than I imagined. From the onset, it was difficult to clearly identify environmental protectors and polluters in the business world. One company, for example, would have an excellent environmental record, but be guilty of human rights offences because of dismal labor standards. Another corporation would have a deplorable environmental history, but would have recently improved their waste management processes due to pressure from consumers and environmental groups. Still others would (unbelievably) show up on both the "most-green" and the "most-toxic" list, such as the popular President's Choice products sold at Canadian grocery stores. It was also extremely difficult to find solid information about corporate environmental practice, except for highly publicized "environmentally-friendly" activities that companies would profile on their websites (such as supporting community 'clean-up' efforts while failing to disclose internal waste records). Due to these factors, and others, I began to think that it would be irresponsible and impossible, to clearly state which companies to support and which to boycott. I started to think that perhaps a paper could be found within these conflicting notions about the need for, and value of, green consumption.

Some environmentalists claim that producers will eventually shift to better resource and waste management because of customer demand and resource scarcity (Hawken, Lovins and Lovins, 1999; Cox, 2004). Others maintain that this will never occur until environmental laws and penalties are implemented and enforced by national governments (Gale, 2002; Hyde, 2005). Still others said that green products were of no consequence and that, instead, the entire system of capitalism should be overthrown in order to stem the damage already created by the corporate world (Maniates, 2002; Princen, Maniates and Conca, 2002). It soon became obvious that the merits of 'green consumerism' are highly contested and required further discussion and analysis.

Process, Products and Patriotism: The History of Consumption

Waste generated as a result of human consumption is currently at its highest point ever and, in 2000, surpassed the earth's natural capacity to absorb it by 15% (Cox, 2004). But this was not always the case. Increased consumption and our propensity for throw-away items began with the industrial revolution. Mass goods were produced cheaply and, as urbanization intensified, cramped living quarters did not allow the space to save items for their reuse (Hyde, 2005). It also became less time consuming to purchase new goods rather than create them by hand. Heather Rogers offers a concrete example of how fast these changes altered the lives of nineteenth century North Americans: "The 1841 version of Catherine Beecher's *Treatise on Domestic Economy* explains how to make candles and soap; her 1869 edition of the same book tells you to just buy those things instead of making them yourself" (Hyde, 2005).

Except for times of war, consumption has been a staple in North American consciousness. Governments and corporations have continuously portrayed consumption as a panacea to whatever troubles were brewing, both domestically and internationally. In post-war periods, and in times of insecurity (such as the months following September 11th in the U.S.), shopping and fervent consumption was promoted as a forum through which citizens could express their nationalism. As Daniel Yankelovich states, "To work hard and consume well was a patriotic duty" (1981, p. 47). These same bodies often reframed consumption as necessary for economic prosperity and domestic stability (Darnovsky, 1996; Durning, 1992; Schor, 1995). For the most part, public opinion reflected these pervasive normative notions: "[M]ost Americans regard [a reduction in consumption] as a threat to their quality of life. More importantly, they are persuaded that the future of their jobs depends on the continuation of current patterns" (Darnovsky, 1996, p. 362). This ideology is exploited by industrialists and economists who state that continuous consumption is necessary, and unquestionable. As capitalism is based on "unfettered access to natural resources" (Rogers as interviewed in Hyde, 2005, para. 18), that "recognizes no limits" (Cox, 2004, para. 13), the planet began to suffer.

Regardless of this reality, there is no indication that "economic growth [will be] ... constrained by the limits of the planet" (Korten, 1996, p. 21). While some citizens call on politicians to enact tougher laws and penalties for corporate negligence, national governments are increasingly constrained by international trade agreements. Many agreements call for sanctions if environmental legislation proposed by any country is thought to represent a barrier to trade (Shrybman, 1999). It is due to this fact, coupled with corporate North America's lack of ability to see beyond the bottom line, that green consumerism has the potential to exert significant pressure on the market and its products. This opinion would be greatly challenged throughout the writing of this paper.

All by Myself: The Individualization of Environmental Responsibility

One of the problems with green consumption, according to its detractors, is that it supports the corporate ideal that places environmental responsibility on the shoulders of individuals.

Many assert that this narrow focus masks larger structures that continue to ensure that wealthy corporations routinely benefit from pollution and the mass extraction of resources. There are facts to support this claim. The instantly recognizable anti-litter slogan and campaign entitled 'Keep America Beautiful' was orchestrated in 1953 by glass, aluminium, paper and steel container manufacturers (Stauber and Rampton, 1995; Berlet and Burke, 1992). These included Pepsi, Coca-Cola, Seagram's, Dupont, Dow, and Procter and Gamble (Darnovsky, 1996; Hyde, 2005). The KAB campaign was created as an alternative to 'bottle bill' legislation that sought to have producers charge a deposit for bottles and cans, repaid to the customer upon their return (Stauber and Rampton, 1995; Berlet and Burke, 1992). In setting up the KAB campaign, these companies effectively moved the environmental onus from those creating the bottles, to those who threw them on the side of the highway. This significant shift has helped to create a culture that looks to individuals to repair the damage caused, in large part, by producers. The role KAB played in creating this context cannot be overstated.

Not even Dr. Seuss is beyond reproach with regards to the "individualization of responsibility" (Maniates, 2002, p. 45). His beloved environmental tale entitled *The Lorax* has been criticized for contributing to the persistent, and corporately supported, notion that individual actions are the most important way to repair ecological damage. In the story, a business owner (called the Once-ler) repents for his greedy and destructive past, and encourages his young friend to plant a tree to symbolize a future commitment to the environment. Michael Maniates claims that the story unintentionally "echoes and amplifies an increasingly dominant, largely American response to the contemporary environmental crisis. This response half-consciously understands environmental degradation to be the result of individual shortcomings (the Once-ler's greed, for example), best countered by action that is staunchly individual" (2002, p. 45). Although it is doubtful that Mr. Maniates holds Dr. Seuss directly responsible for such a pervasive worldview, his point is well illustrated and well taken.

Maniates later states that green consumption threatens 'real' environmental activism, and detracts from larger corporate structures that continue to go unchallenged:

> This collective obsessing over an array of "green consumption" choices ... is noisy and vigorous, and thus comes to resemble the foundations of meaningful social action. But it is not, not in any real and lasting way that might alter institutional arrangements and make possible radically new ways of living that seem required (2002, p. 52).

Finally, he states that people must reclaim their citizenship by placing civic involvement before consumption: "Confronting the consumption problem demands, after all, the sort of institutional thinking that the individualization of responsibility patently undermines. It calls too for individuals to understand themselves as citizens in a participatory democracy first, working together to change broader policy and larger social institutions, and as consumers second" (Maniates, 2002, p. 47). These are convincing and purposeful arguments that challenged my premise: save the world through shopping?

In further reading, I was soon heartened to find researchers who made claims that people participate in the civic and market arena in different ways. Who is to say that my decision not to buy paper products harvested from virgin forests (and, perhaps, sending a letter to the company articulating this choice) is less significant than someone who publicly protests the way in which the federal government supports this same company? Some might argue that the second choice is more valuable as it may heighten other people's awareness of the company's practices. But, in exercising my choice, I am speaking to the corporation through the representation of money, which many claim is the only dialect that companies understand. Even if I am not willing to appear on the national news for an environmental cause, I can still create social change.

Can product choice be reframed as another arena for participatory democracy? Jeremy Rifkin (1990) states that consumption is a perfect venue for individuals to begin to exert corporate power: "By extending the concept of participatory democracy [to consumption] ... we assure our responsibilities as active participants in the decisions that affect the future course of our society, civilization, and ultimately, the planet" (p. xv).

Elkington, Hailes and Makower (1990) illustrate how the conventions of capitalism can work to the consumer's advantage:

> You may be surprised at how easy it is to make your voice heard in the marketplace. The marketplace is not a democracy; you don't need majority opinion to make change. Indeed, it takes only a fairly small portion of shoppers—as few as one person in ten—changing buying habits for companies to stand up and take notice (p. 9-10).

Fred Gale (2002) notes that, "cautious consumers can affect dramatic change through their purchasing power" (p. 299), and Brecher, Costello and Smith (2000) encourage citizens to "utilize the power that lies hidden in the withdrawal of consent" (p. 31). Admittedly, some of the voices that champion green consumerism also peddle books about the various ways that consumers can heal the earth; but does this mean that their contributions should be discounted? If I try to create an environmental business, does the fact that I am seeking to make a living negate my environmental concerns and suggestions? This leads to another main criticism of green consumerism: that it only serves to solidify capitalism, thus perpetuating continuous ecological depletion, and reinforcing market inequities.

Stay the Course? Capitalism and Green Consumption

In *The Business of Consumption*, Laura Westra (1998) sums up the green consumerism critique quite accurately when she writes: "It is not about *how* to conduct business but about *whether* to continue to sustain an enterprise that is based on increased consumption" (p. ix). Some environmentalists see green consumerism as reinforcing capitalism, as it offers products and processes which may mitigate some ecological effects, but which do not question the 'assumption of consumption'. As Princen, Maniates and Conca (2002) state:

> Consumption becomes sacrosanct. If water supplies are tight, one must produce more water, not consume less. If

> toxics accumulate, one must produce with fewer by-products —or, even better, produce a cleanup technology—rather than forego the production itself. Goods are good and more goods are better. Wastes may be bad—but when they are, more productive efficiencies, including eco-efficiencies and recycling, are the answer. Production reigns supreme because consumption is beyond scrutiny (p. 5).

These critics also point to the ways in which a green sensibility has been co-opted by advertising and marketing executives; a trend that the business community admits and celebrates (Darnovsky, 1996). Michael Maniates notes that capitalism possesses the "dynamic ability" to market dissent, and then "sell it back to dissenters" (2002, pp. 46 and 51). He states that this has contributed to the boon of green industry which becomes "paradoxically, a consumer-product growth industry" (Maniates, 2002, p. 47). Again, this criticism is well taken. But the fact remains; regardless of ideology or personal politics, we continue to live in a consumer society.

While Marcy Darnovsky agrees that "commodifying dissent has become standard marketing practice" (1996, p. 161), she sees some possibilities for subversion. She presents a new term for conscious shoppers when she writes "[c]ritics of consumer capitalism often argue that it substitutes consumer culture for political involvement. The activist-consumer challenges this assumption" (Darnovsky, 1996, p. 161). The activist-consumer is "simultaneously a marketing category and a social identity" (Darnovsky, 1996, pp. 160-161). They are also "a major segment of the consuming public" (Dadd and Carothers, 1990, p. 13). Although capitalism is undoubtedly the cause of many social and environmental ailments, it is still the system of choice for many North Americans. Is there potential for green action within the confines of this structure?

In the book *Natural Capitalism*, Paul Hawken and Amory and Hunter Lovins (1999) write about the futility of attempting to retreat from capitalism, and the hidden environmental power it may actually possess. They write:

> A common response to the misuse, abuse, or misdirection of market forces is to call for a retreat from capitalism and a

> return to heavy-handed regulation. But in addressing these problems, natural capitalism ... suggest[s] that we should vigorously employ markets for their proper purpose as a tool for solving the problems we face (p. 260).

'Natural capitalism' is defined as "a set of trends and economic reforms that reward energy and material efficiency", and as a "coherent theory of how to exploit market systems... to generally support the goals of environmentalism" (Labour-Talk-Law Online Encyclopedia, March 2005). Green consumerism falls within this definition. As *Natural Capitalism* reminds us, the goal and 'true purpose' of the market in a capitalist system is to "allocate scarce resources efficiently over the short term" (Hawken, Lovins and Lovins, 1999, p. 261). This will become a more valuable trait as resource extraction leads to scarcity, and the need for new alternatives. Hawken, Lovins and Lovins echo others when they suggest that "[c]ompanies that ignore the message of natural capitalism do so at their peril" (1999, p. xiii). This is due not only to depleting natural resource pools, but also because of a more conscientious consuming public.

As corporations are beginning to understand the power wielded by the green consumer, they have attempted to 'green up' the presentation of their products and services. This has led many environmental groups to investigate the claims of companies, often with disappointing results.

Green on the Outside: The Practice of Corporate Greenwashing

Greenpeace coined the term 'greenwashing' in 1991 to "describe corporate deceptions, exaggerations, and cosmetic changes undertaken purely as public relations ploys" (Darnovsky, 1996, p. 223). These "cleverly disguised attempts" (Freeman, Pierce and Dodd, 1998, p. 345) often extol a virtuous environmental record, or donate money to green organizations to draw attention away from questionable ecological activities. Another type of greenwashing occurs when companies advertise their 'environmental-friendly' products or processes, while failing to share that these standards are the minimum required by law

(Hyde, 2005). The 'Keep America Beautiful' campaign discussed earlier is generally thought to be one of the first examples of corporate greenwashing (Hyde, 2005). As a result, many environmentalists believe that any attempt to place the blame for environmental destruction on individuals feeds into this phenomenon (Darnovsky, 1996).

Having considered the possibility that many companies I support may have inflated, if not invented, environmental practices, I am still wary of throwing the baby out with the bathwater. Maybe hope resides in individual action. Statistics Canada states that, in 2003, nine of ten Canadians polled listed the environment as one of their top concerns (Boyd, 2004, p. 1). The law of averages states that there must have been some business owners among those polled. As much as corporations may have profit as their guiding principle, many individuals have their children's health as their main priority. It is reasonable to assume that some of that concern translates into corporate environmental responsibility. As Marcy Darnovsky notes "Four kinds of problems beset green marketing: criticism from environmentalists, consumer scepticism, an uncertain and chaotic regulatory situation, and the difficulties inherent in evaluating what makes a product, package or process 'greener'" (1996, p. 250). Which leads to the question—what role does government play in maintaining the environmental standards of corporations?

Minding the Store: Government and Green Consumerism

It would seem reasonable for the argument to be made that government regulation needs to be a part of the green consumerism movement. How else can we (the consumer) believe some of the claims that advertisers make about the biodegradability of their dish soap, or the organic origins of their cotton? But it is important to remember, as Heather Rogers notes, that "[t]he state helped create the disposable society we have today" (Hyde, 2005, para. 20). By consistently intervening to subsidize polluting activities, and by helping to keep the cost of natural resources artificially low, governments have often served to impede environmental activities, rather than enhance them (Hyde, 2005; Hawken, Lovins and Lovins,

1999). In fact, some environmental writers and thinkers are extremely critical of the ways in which a 'sustainable sensibility' has been co-opted by mainstream political parties and 'liberal' environmentalism. The latter has been dismissed as being "so compatible with contemporary material and cultural currents that it implicitly supports the very things it should be criticizing" (Wapner, 1996, p. 21). Princen, Maniates and Conca agree with this assertion, when they write, "[t]he environmental movement is very middle class ... and its organizations do not challenge middle class values" (2002, p. 8). These statements articulate a concern that any adoption of 'radical' principles by mainstream organizations, and governments, only serves to dilute the message and intent they were originally imbued with.

Despite this belief about the inability of governments to address ecological issues, these same voices chide the average citizen for their lack of political advocacy. Maniates (2002) states "When confronted with environmental ills—ills many confess to caring deeply about—Americans seem capable of understanding themselves almost solely as consumers who must buy "environmentally sound" products (and then recycle them), rather than as citizens who might come together and develop political clout sufficient to alter institutional arrangements that drive a pervasive consumerism" (p. 51). This is a troubling and contradictory statement given the previous arguments. If environmental academics have no faith in government and the environmental movement, than why should John and Jane Q. Public? Shouldn't any environmental effort, which may lead to increased ecological awareness, be celebrated?

Thankfully, other writers posit their calls for action in more positive terms, citing corporate culture as the issue needing address: "We can't just leave this up to industry, and we can't just leave this up to the market—we need to intervene because there's a fundamental lack of democracy in the use of our resources" (Rogers as interviewed in Hyde, 2005, para. 26). One of the most meaningful ways for the public to converse with industry is by altering their shopping patterns. This is an easier practice for the wealthier segments of society to undertake, for obvious reasons (access to transportation, ability to purchase more expensive products, time to comparison-shop, etc.). The goal of this exercise is emphasized when communities from all socio-economic stratums decide to elect government officials who cite the environment as an central issue, and whose plans include concrete processes to encourage and enforce ecological compliance among industry. This requires *more* government intervention, guided by public concerns.

One way for this to happen is for governments to gradually increase the prices of natural resources, until they begin to reflect their full cost. 'Full cost' means a price that not only illustrates the cost of extraction and the scarcity of the resource, but also accounts for disposal and biological breakdown of the product (Hawken, Lovins and Lovins, 1999). Although this causes short-term financial strain for consumers (i.e. gas prices in North America which, although climbing, are still artificially low compared to other parts of the world), it does encourage conservation and responsible production. Paul Hawken states that 'full cost' economics has other benefits as well, such as lessening the divide between rich and poor, but requires that governments, and citizens, make decisions that are difficult and unpopular: "Unless we take business out of politics, prices will never reflect cost or value. Unless it is in our best interest to live well within our means, income polarization will broaden" (Hawken as interviewed in Cox, 2004, Economy Class section, para. 3). This is also a challenge for green businesses whose products are usually priced higher than market value. Although this may give rise to arguments that only the rich can afford to live sustainably (and use the remainder of naturally-occurring resources), the hope is that full cost pricing will necessitate cheaper, more ecologically-sound alternatives and, again, encourage conservation.[1]

North American governments can also look to other countries that have implemented successful environmental programs. In Germany, the state decreed that 72% of the bottles that a company produces be refillable (Hyde, 2005). In Denmark, 98% of the bottles are refillable, and 98% of the surveyed public stated that they consistently return the bottles for a deposit (Hyde, 2005). By mandating company practices, these countries' governments have significantly altered the environmental activities of the general public.

Another possibility is government certified 'eco-labeling' for products that are proven to exert less of a strain on the natural environment. Again, the argument can be made that this does not address larger issues of North American consumption and can thus only result in "small incremental changes in product and production environmental standards" (West, 1995, p. 20). However, it is undeniable that eco-labeling may serve to "make the same choice available to the interested-but-not-fully-committed consumer who wants to 'do the right thing' but not at great personal inconvenience" (Gale, 2002, p. 298). Fred Gale (2002) states, "This larger segment of consumers constitutes the mass of shallow ecological purchasers that might yet be encouraged to achieve deeper ecological objectives" (p. 298). When this 'larger segment of consumers' decides to make 'small incremental' modifications to their shopping patterns, large changes can result. Government has a significant role to play in setting the environmental standards to which any product baring an eco-label must adhere.

A responsive and proactive government must work in collaboration with an empowered citizenry to ensure that an environmental ethic begins to infiltrate and alter the market. And although the mainstream environmental movement may be flawed, Darnovsky notes that its central values offer an alternative worldview that has implications for issues other than ecology: "[T]he environmental movement seems to me our best hope for popularizing global political awareness and a willingness to rethink not just consumer culture, but all the fundamental arrangements of the political and economic order" (1996, p. 374). This consciousness that encourages individual 'rethinking' also contains possibilities for larger, cohesive actions.

Not Shopping Together: Collective Action and Social Change

When discussing issues pertaining to environmental choices, it is helpful to view consumption as more than "an individual's choice among goods. [It is also] a stream of choices and decisions winding its way through the various stages of extraction, manufacture, and final use, embedded at every step in social relations of power and authority" (Princen, Maniates and Conca, 2002, p. 12). Shopping for products can be a "significant part of an individual's attempt to find meaning, status, and identity" (Princen, Maniates and Conca, 2002, p. 14). If this act is politicized, and understood to have corporate, social and environmental consequences, these individual choices can create a space for contemplation, deliberation and caution. As Brecher, Costello and Smith note, this can then become an arena where people discuss their everyday decisions, and begin to come together:

> This process may start with some people internally questioning or rejecting some aspects of the status quo [such as mass consumption at Christmas, or buying products from environmentally irresponsible companies]. It becomes a social process as people discover that others are ... asking the same questions, and being tempted to make the same rejections.... Seeing other people share similar experiences, perceptions and feelings opens up a new set of possibilities. Perhaps collectively we can act in ways that have impacts isolated individuals could never dream of having alone. And if we feel this way, perhaps others do, too (2000, p. 20).

As Kalle Lasn, one of the founders of *Adbusters* states "Individual change *and* collective action geared to enforcing corporate responsibility is the *only* way that we will together achieve our goals" (Maniates, 2002, p. 215). The goal of environmental awareness, conservation and protection is within reach, and green consumption is a venue where individual responsibility can meet communal activism.

Everyday Activism: Green Consumerism as Environmental Awareness

Brecher, Costello, and Smith (2000) state that civic engagement is powerful, but is manufactured to seem elusive: "The latent power of the people is forgotten, both because those in power have every reason to suppress its knowledge and because it seems to conflict with everyday experiences in normal times" (p. 23). An encouraging component of green consumerism is that it is mired in the everyday. In his book *Reveille for Radicals* (1969), Saul Alinsky writes, "Most people are eagerly groping for ... some way in which they can bridge the gap between their morals and their practices" (p. 94). Green consumerism offers this possibility because, regardless of shopping habits, everyone must buy food. And, as Professor Jules Pretty of Essex University notes, "The most political act we do on a daily basis is to eat" (Connor, 2005, para. 10). Because "global food production and trade ... consumes more fossil fuel than any other industrial sector" (Shrybman, 1999, p. 12), this section of the market offers a variety of ways for 'cautious consumers' to greatly lessen their ecological impact. By choosing to buy local produce, shoppers are significantly reducing the 'food miles' between the farm and their plates, resulting in less fossil fuel use, fresher produce and more money for their neighborhood businesses. This small change can be a catalyst for becoming more ecologically-mindful while shopping.

Following the Green: Industry Trends

Perhaps one of the most promising aspects of green consumerism is that improvement is imminent. Hawken, Lovins and Lovins (1999) write of the next "industrial revolution", where, aided by 'full cost' economics and government regulated taxation on waste output, producers begin to look for the most efficient and sustainable way to conduct business. They state that the waste and noise associated with current manufacturing practices "represent money being thrown away. They will disappear as surely as did the manure from the nineteenth-century streets of London and New York. Inevitably, industry will redesign everything it makes and does, in order to participate in the coming productivity revolution" (Hawken, Lovins and Lovins, 1999, p. 13). And, many environmentalists assert, these producers need look no further than the insect in front of them to discover unparalleled proficiency, productivity and design. When comparing man-made Kevlar (a strong fiber) to the silk spun by a spider, a clear winner emerges:

> The spider manages to make equally strong and much tougher fiber at body temperature, without high pressures, heat or corrosive acids ... If we could learn to do what the spider does, we could take a soluble raw material that is infinitely renewable and make a superstrong water-insoluble fiber with negligible energy inputs and no toxic outputs (Benyus, 1997, p. 135).

As science writer Janine Benyus notes, "We don't need to invent a sustainable world—that's been done already" (1998, keynote address). If scientists and producers could work together to create sustainable processes that mimic those found in nature, the possibilities are limitless. With such forward thinking, and such fine examples, green consumption has the opportunity to make zero waste the goal, and then the standard, for the production of all goods.

The Social Worker Stands Back

A large reason why environmental awareness is important for the field of social work is because of the people who are first impacted by ecological deterioration: the poor, people of color, women, people of the global south, etc.[2] It is interesting, then, to write an entire paper which is largely based on choices surrounding money and whether it can be spent 'greenly' or not. Many of the criticisms of green consumerism are valid. It is not a panacea for all environmental ills, and larger issues around the creation of needs and mass over-consumption must be addressed. But telling people to buy less is too simple an answer. As Juliet Schor writes:

> In a society in which consumption is structurally positioned as the answer to so many needs, desires, and problems, and in which alternatives are structurally blocked, moral suasion is insufficient. Asking people to act ethically is important,

but we must also analyze and transform the structures that make it difficult for them to do so (1995, p. 21).

In lieu of this, it is vital to provide people with the information that they may not have the ability or inclination to access. What products and services are available locally and what are some of the best environmental practices that other individuals have been able to implement?

Talking and writing about the environment is the first step towards action, and should not be relegated to those with more expertise or experience. As Frank Cross asserts, "Just as wars are too important to leave to the military, the environment is too important to leave to the environmentalists" (1990, p. 46).

Conclusion

The debate will go on. Some environmentalists will continue to dismiss green consumption as the latest market trend, resulting in marginal environmental benefits. Others will claim that green consumption is part of the problem of capitalism and should thus be dismissed as contributing to a system that necessitates environmental damage. But others, like me, will say that the game of sustainable living begins when more people can play. And anything that encourages greater contemplation of, and participation in, green issues is worth examining.

As Princen, Maniates and Conca state, "Ultimately, the challenge is not just to confront consumption but to transform the structures that sustain it" (2002, p. 328). But *how* do we transform the structures of consumption? We start by taking our time when we shop; looking at where things are made, where human rights are marginalized, how products are packaged, and what they contain. We think about buying less and paying more for products that we feel good about purchasing. We talk to other people about where they get their produce, how they keep their energy bills down, and why they don't have a car. We plan parties that are plastic free, and ask for less for Christmas. We become educated about current global initiatives, such as the United Nations Environment Program (http://www.unep.org) and the Millennium Development Goals (http://www.un.org/millenniumgoals). We elect leaders that represent our environmental values and insist that they talk to other governments about how best to preserve and protect the planet.

When those who can afford to begin to ask for products and programs that foster environmental responsibility, the laws of the marketplace dictate that more affordable environmental goods and more widespread initiatives will follow. Small, incremental changes may be derided as inconsequential, but they are perhaps the only way that substantial social change begins. And it is we, who posses the wealth and the resources, who have the most work to do.

Appendix: 12 Ways to Simplify Your Life and Save the World

1. Avoid shopping
2. Leave the car parked
3. Live in a nice neighborhood (that will allow you to walk to stores or easily access public transport)
4. Get rid of your lawn
5. Cut down on your laundry
6. Block junk mail
7. Turn off the TV
8. Communicate by email
9. Don't use a cellular phone
10. Drink water rather than store-bought beverages
11. Patronize your public library
12. Limit the size of your family.

Endnotes

1. The same argument is often made about how-to lists such as *12 Things You Can Do To Save the Earth* (please see Appendix); namely that they that presuppose an affluent lifestyle. My response to this argument is that small environmental changes, and lists instructing how to make them, should be celebrated for what they are: a beginning for car owners and lawn tenders. If we claim the middle class are among the greatest polluters, than we should not dismiss resources that encourage 'green' activities, however, small.

2. Please see John Coates (2003) and Maria Mies and Vandana Shiva (1993) for a more comprehensive analysis.

References

Alinsky, S. (1969). *Reveille for Radicals*. New York: Vintage Books.

Benyus, J.M. (1997). *Biomimicry: Innovations Inspired by Nature*. New York: William Morrow.

Benyus, J.M. (1998, October 8). Keynote Address. *E-Source Members Forum*. Aspen, Colorado. Retrieved March 15, 2005, from http://www.esource.com

Berlet, C. and Burke, W.K. (1992). "Corporate Fronts: Inside the Anti-Environmental Movement", *Greenpeace Magazine*, January/February/March 1992, pp. 8-12.

Boyd, D.S. (2004). *Sustainability within a Generation: Executive Summary*. Vancouver: David Suzuki Foundation.

Brecher, J., Costello, T. and Smith, B. (2000). *Globalization from Below: The Power of Solidarity*. Cambridge, Massachusetts: South End Press.

Coates, J. (2003). *Ecology and Social Work: Toward a New Paradigm*. Halifax: Fernwood Publishing.

Connor, S. (2005, March 3). "Buy Local Produce and Save the World: Why Food Costs £4bn more than We Think", *The Independent (London)*. Retrieved March 5, 2005, from http://www.findarticles.com/p/articles/mi_qn4158/is_20050303/ai_n11845617

Cox, S. (2004, April 27). "From Here to Economy", *Grist Magazine*. Independent Media Institute. Retrieved April 11, 2006, from http://www.alternet.org/story/18518

Cross, F.B. (1990). "The Weaning of the Green: Environmentalism Comes of Age in the 1990s", *Business Horizons*, September/October 1990, pp. 40-46.

Dadd, D.L. and Carothers, A. (1990). "A Bill of Goods? Green Consuming in Perspective", in C. Plant and J. Plant (eds.) *Green Business: Hope or Hoax*. Santa Cruz, California: New Society Publishers.

Darnovsky, M. (1996). *The Green Challenge to Consumer Culture: The Movement, the Marketers, and the Environmental Imagination*. Doctoral dissertation, University of California, Santa Cruz.

Durning, A. (1992). *How Much is Enough? The Consumer Society and the Future of the Earth*. London: Worldwatch Institute.

Elkington, J., Hailes, J. and Makower, J. (1990). *The Green Consumer*. New York: Penguin Books.

Freeman, R.E., Pierce, J. and Dodd, R. (1998). "Shades of Green: Business, Ethics, and the Environment", in L. Westra and P.H.

Werhane (eds.), *The Business of Consumption: Environmental Ethics and the Global Economy*. Maryland: Rowman and Littlefield Publishers, pp. 339-354.

Gale, F. (2002). "Caveat Certificatum: The Case of Forest Certification", in T. Princen, M. Maniates and K. Conca (eds.), *Confronting Consumption*. Cambridge, Massachusetts: The MIT Press, pp. 275-300.

Hawken, P. Lovins, A. and Lovins, L.H. (1999). *Natural Capitalism: Creating the Next Industrial Revolution*. Boston: Little, Brown and Company.

Hyde, A.C. (2005, April 4). "Litterbug World", [Interview with Heather Rogers, Filmmaker of Gone tomorrow: The hidden life of garbage]. *LiP Magazine*. Retrieved April 11, 2006, from http://www.alternet.org/story/21651.

Korten, D.C. (1996). "The Failure of Bretton Woods", in J. Mander and E. Goldsmith (eds.), *The Case Against the Global Economy*. San Francisco: Sierra Club Books, pp. 20-30.

Labour-Talk-Law Online Encyclopedia. (2005, March). Retrieved March 25, 2005, from http://encyclopedia.laborlawtalk.com/Natural%20Capitalism

Maniates, M. (2002). "Individualization: Plant a Tree, Buy a Bike, Save the World?" in T. Princen, M. Maniates and K. Conca (eds.), *Confronting Consumption*. Cambridge, Massachusetts: The MIT Press, pp. 43-66.

Maniates, M. (2002). "In Search of Consumptive Resistance: The Voluntary Simplicity Movement", in T. Princen, M. Maniates and K. Conca (eds.), *Confronting Consumption*. Cambridge, Massachusetts: The MIT Press, pp. 199-236.

Mies, M. and Shiva, V. (1993). *Ecofeminism*. Halifax: Fernwood Publications.

Peattie, K. (1992). *Green Marketing*. London: Pitman Publishing.

Princen, T., Maniates, M. and Conca, K. (2002). "Confronting Consumption", in T. Princen, M. Maniates and K. Conca (eds.), *Confronting Consumption*. Cambridge, Massachusetts: The MIT Press, pp. 1-20.

Princen, T., Maniates, M. and Conca, K. (2002). "Conclusion: To Confront Consumption", in T. Princen, M. Maniates and K. Conca (eds.), *Confronting Consumption*. Cambridge, Massachusetts: The MIT Press, pp. 317-328.

Rifkin, J. (ed.). (1990). *The Green Lifestyle Handbook: 1001 Ways You can Heal the Earth*. New York: Henry Holt and Company.

Schor, J.B. (1995). "A New Economic Critique of Consumer Society", *Report from the Institute for Philosophy and Public Policy, 15* (4): 17-21.

Shrybman, S. (1999). *The World Trade Organization: A Citizen's Guide*. Ottawa, Ontario: Canadian Centre for Policy Alternatives.

Stauber, J. and Rampton, S. (1995). *Toxic Sludge is Good for You: Lies, Damn Lies and the Public Relations Industry*. Monroe, Maine: Common Courage Press.

Seuss, Dr. (1971). *The Lorax*. New York: Random House.

Wapner, P. (1996). "Toward a Meaningful Ecological Politics", *Tikkun,* 11 (3): 21-22.

West, K. (1995). "Ecolabels: The Industrialization of Environmental Standards", *The Ecologist,* 25 (1): 16-20.

Westra, L. (1998). "Preface", in L. Westra and P.H. Werhane (eds.), *The Business of Consumption: Environmental Ethics and the Global Economy*. Maryland: Rowman and Littlefield Publishers, pp. ix-x.

Yankelovich, D. (1981). *New Rules: Searching for Self-fulfillment in a World Turned Upside Down*. New York: Random House.

The Ecological Explanation for the Environmental Crisis

3

Ryder W. Miller

The environmental crisis requires our attention like it has never done before. The year 2004 was a breakthrough year where it was widely acknowledged that we need to be concerned about climate changes. Finally, even some of the critics are now convinced that it is real. Icebergs are adrift, the global air temperature has increased, and water levels are rising. Eight of the last ten years have been the hottest recorded in human history. Polar bears are drowning because the distances between ice islands have gotten too large. In some areas glaciers have disappeared. Simulation photographs in magazines show some of our major cities under water. Also, harmful non-indigenous species are replacing native species, the rainforest and rare habitat are being destroyed, pollution legislation is being defanged, and the environmental movement could not rally enough support in 2004 to replace George W. Bush.

Out of necessity, the environmental movement needed to renew itself, to finds its way out of the rut it found itself in the aftermath of the 2004 election. The historical arguments and pronouncements of the environmental movement are still easy to understand and justify. The root causes of the environmental crisis can be explained by understanding roughly 20 ecological concepts. Environmentalism grew out of ecology and the preservation movement, which began by focusing on the protection of rare species and rare habitats. Ecology has blossomed into a science and a set of related concerns encapsulated by the term environmentalism. But one can find the root explanation of the environmental crisis by searching its ecological roots. Some of the biggest environmental (ecological) problems include global warming, pollution poisoning, extinction, harmful non-indigenous species, habitat destruction, and overpopulation. There are historical ecological concepts that explain what caused these problems.

But first, what is ecology? Ecology among other things is a natural science. Ecology means a number of things to different people, but its scientific definition is the study of the interaction between organisms and their environment. It is also concerned with the study of the interaction between organisms. There are competition ecologists who study the competition within the same species and between different species. There are niche ecologists who study how specific organisms make their living in their environment. The science of ecology is also concerned with large environmental chemical processes like oxygen, nitrogen, and water cycles.

Science is a methodology to ask questions and gather information to answer them. Science is a perspective with a set of values, assumptions, historical precedents, and lessons. Ecology is a new science which seeks to document and answer questions about connections in the natural world. Its concepts and ideas can explain the mess that we are now in. With the knowledge of a small number of ecological concepts one can explain the causes of the major environmental issues we face today.

What follows are the ecological explanations for six severe environmental problems.

1. Global Warming

Global warming is the result of societal overproduction of greenhouse gasses: mostly carbon dioxide but also others. Greenhouse gasses create a barrier in the atmosphere that traps heat near the surface of the planet. Just like in a greenhouse, cosmic heat can enter the atmosphere of the planet, but greenhouse gasses, just like glassing in a greenhouse, creating a barrier that keeps the heat from dispersing into space. Without greenhouse gasses the planet would be as cold as the vacuum of space. With too much greenhouse gasses the planet can become overheated like Venus where the surface temperature is hot enough to melt lead. We need greenhouse gasses to survive, but our overuse of fossil fuels has resulted in the creation of excess greenhouse gasses that will result in more heat being retained in the atmosphere. As a result, the temperature of the planet is rising. Over the past one hundred years the Earth's temperature is estimated to have risen between 0.5 and 1.25 degrees Celsius. If the current production of greenhouse gasses persists it is estimated that by the year 2100 the temperature of the Earth will rise between 1.8 and 5.8 degrees Celsius (Slobodkin, 2003).

The amount of carbon dioxide in the atmosphere is estimated to have increased by 30% in the last century and a half. A minor component, carbon dioxide is only slightly more than three parts per hundred thousand in the atmosphere (Slobodkin, 2003). When fossil fuels are broken down to provide energy for our millions of automobiles, excess carbon dioxide is produced and enters the atmosphere. Carbon dioxide is also produced by respiration, the process by which we process the oxygen, carbon, and water we need. We breathe in oxygen and exhale carbon dioxide. Plants consume carbon dioxide through the process of photosynthesis whereby cosmic energy from the sun is used to create the chemical bonds that form the energy rich sugars from carbon dioxide and water. Animals eat plants and gain energy from these compounds that are fueled by cosmic energy from the sun. When the compounds are broken down or digested, the animals gain the energy they need to survive. The carbon and oxygen cycle processes cosmic energy which is transferred between plants and animals through the processes of photosynthesis and respiration, but industrialism has produced excess carbon dioxide and other gasses which affect the whole cycle by heating the whole global system. The result will be raising water levels due to water expansion caused by rising temperatures and an increased ocean volume caused by the melting of the polar ice caps. This increase in water volume will affect chemical cycles that affect all living things on the planet. Life has become an integral part of these chemical cycles that they depend upon. The deforestation of the rainforests also results in less carbon dioxide being absorbed by living systems. The ocean will only absorb a portion of the excess carbon dioxide in the process causing deleterious decalcification of marine life by making the ocean more acidic.

Plants are the base of the food chain that converts chemicals from the atmosphere and nutrients from the soil to produce the energy stocks that fuel the chain of life. Plants need chemicals such as phosphorus and nitrogen, as well as water and carbon dioxide, which also have global flow processes that include transport between the atmosphere, water, life, and soil. We can impact these cycles by the production of chemical pollutants. When we release excess chemicals into the atmosphere or bodies of water we shift the natural cycles that can result in the demise of organisms that have depended upon these sensitive global cycles.

Red tides are the result of excess nutrients entering water systems that can result in the deleterious success of microscopic life. With excess nutrients these microscopic animals can grow beyond their carrying capacity, i.e. the population level that the natural environment can support. A population in excess, even microscopic, will deplete the natural environment of its requirements. This can create exhausted or dead water systems that undermine all the creatures higher up on the food chain. Such situations are akin to knocking the legs out from underneath a table. The success of our species has the rest of the natural world trying to adapt to the changes we have caused. The excess carbon dioxide we have produced is heating up the whole planet. If left unchecked, more areas of the planet are likely to become unlivable.

We have also seen the more immediate effects of other dangerous chemicals we have let out into the environment.

2. Pollution Poisoning

The chemicals that enter the biosphere of living systems as a result of industrial processes are not all beneficial or natural, i.e. they do not belong. Plants need carbon, nitrogen, etc., but there are many things that we have introduced into the world through industrial processes that are harmful to ourselves and other living creatures. We have produced dangerous chemicals that enter living systems and accumulate through the process of biomagnification, an increase in concentrations in living tissue as one travels up the food chain.

Plants (organisms that can photosynthesize, thereby gaining energy from sunlight) are consumed higher up the food chain by animals that cannot photosynthesize, and animals higher up the chain then eat these animals. At the top of the chain is the apex predator that serves as an indicator species for the whole ecosystem: an interactive collection of numerous creatures. Food webs are more complicated than food chains in that the relationships are not linear, but rather a collection of many interconnections of food chains. The creatures that eat photosynthesizing bacteria may be eaten by many creatures, but these critters may also help in the digestion of larger animals which are connected in webs of relationships rather than a linear chain upwards.

Pollution enters these webs and chains at all levels and the concentrations increase up the many chains, until it could pose a threat to the animals at the top of the chain. Human society is often at the top of the food chain. The animals we eat can potentially pollute us. We have learned that what we have polluted the environment with can wind up on our dinner plate due to biomagnification. These pollutants can cause problems and cancer in human beings as well as creatures in the wild. For example, DDT had a deleterious effect on brown pelicans by interfering with the production of their eggs, which were too thin to incubate their young. When the use of DDT was diminished the brown pelicans made a recovery. Pollution is also one of the factors that lead to the demise of wildlife populations. Through natural cycles we have all encountered and ingested dangerous chemicals. We need to wash the fruit we find at the supermarket because of pesticides that are used to deter insects that would eat what we grow. Birds eat these poisoned insects and are poisoned. Excess pesticides wind up in oceanic and riparian (river system) ecosystems, flowing downstream. Pollutants have also been dumped into bodies of water resulting in a poisonous harvest.

We as well as the other organisms depend upon these complicated cycles. When they are disturbed things disappear.

3. Extinction

Wild, rare animals have been compared to priceless pieces of art (Wilson, 1984; Slobodkin, 2003). They have developed over millions of years through the process of evolution, which is really easy to explain and understand despite the intelligent design controversy. It is surprising why the belief in evolution has been so controversial. Theological arguments seem superfluous (Futuyma, 1984) or unnecessary when one understands the evolutionary process at work. Evolution can be explained by the following set of propositions or principles.

Diversity or differences exist within populations (members of the same species) of animals. Some of us are tall, others short. Some fat, others skinny. There is continuum of skin colors and hair colors. Some of these differences are caused by genetics, i.e. we have a different genetic makeup. Even though we are all members of the same species, i.e. we are similar and together we can reproduce more of our kind, we are different from each other.

One of the factors that create these differences between us is caused by natural mutations of our genetic material. We pass our genes (biological building instructions) from generation to generation, but sometimes this biological information is changed between generations creating more differences between members of a species. Species in groups also break apart and change so they can adapt to different local environments and situations. Some of these differences between us are more advantageous than others and are therefore more likely to be selected in the process of reproduction or sexual selection. When times are barbaric it may be more advantageous to have large children because they are more likely to survive battles. Birds of prey, like other animals, through natural selection are selected for features that enhance their ability to hunt. Natural selection is the process by which organisms compete to shape future generations. Local conditions and mutations result in differences, some of which are more beneficial to the species or competitive than others. The process results in different parts or groups of a species growing separate from each other. Sometimes the various populations of different species have grown so far apart that they cannot reproduce between each other, i.e. they cease to be part of the same species. Natural selection over billions of years has created the natural world which seems miraculous to many. The competition can be so fierce that species need to change over generations to survive in their niche or place and role in the environment. An animal or plant's niche is their way of making a living, a role that they have evolved to fill. That the fierce competition of natural selection should may be acknowledged but not emulated is part of driving force behind the arguments of intelligent design, but such proponents have yet to present a competing theory with scientific merit.

Living things are the result of million of years of evolution. As things change some species die off and others emerge. But we have the consciousness and the power to protect the other living things on the planet. We can mass-produce tools, we can pass huge amounts information from generation to generation; we can document the damage we have caused the environment. We can adjust to save these creatures that have evolved through the same process we have. Having understood the process that created ourselves we also have the potential to grow beyond them. We can thrive through cooperation. The presences of nuclear weapons, which threaten our very existence, require that we do so. But the changes we make to ourselves during our life and the decisions we make will not impact the genes we pass to our offspring. Natural selection passes along the genetic heritage rather than how it was expressed under the environmental conditions in which organisms develop. When we observe wildlife we are watching the interaction between genetic determination and impacted expression. But not all organisms have survived.

Extinction means that creatures have disappeared from the entire face of the planet. It means they are gone and will never exist again. The passenger pigeons that once flew across areas of America in flocks so big that they blocked out the entire sky were eradicated from the continent by hunting and habitat destruction. Extirpation, similar to extinction, means that a species had been removed from certain locations. One may be able to find them elsewhere, but from certain locations they are now gone, usually resulting from human actions.

When we change the environment we alter the fragile connections that exist. Creatures have depended upon these connections that we have altered. We need to keep an eye on our neighbors on the planet to make sure they do not disappear. Newcomers also pose a problem.

4. Harmful Non-Indigenous Species

Newcomers, even if only non-human animals and plants, have the potential to change and alter all established ecosystems, i.e. an assemblage of weather-dependent connections and relationships between organisms. Harmful non-indigenous creatures do not have the population limiting factors that have kept them in check in their original locations from which they are. There is a not a population of predators which depend upon them as a food source. There are not pathogens that are adapted to living within their populations. Without these population limiting factors, non-indigenous species out-compete native species. In the process of out-competing these native or previously resident creatures, the environment is altered. Species can be extirpated because of the success of these newcomers.

Harmful non-indigenous species are a problem on the same level as habitat destruction in its detrimental effects on native creatures. Local ecosystems are overrun with non-native species all across the world. There is a movement to return habitats, the biological communities in which animals live, back to their native conditions, i.e. before the advent of the biological newcomers. This movement takes the form of the habitat restoration movement that usually involves the removal of non-native plants and the propagation of native plants. These actions can create more diversity and therefore homes for a greater variety of animal and plants. But many people have favorite non-native species that they think should be an exception in the recreation of the past. For example, urban birds like pigeons, starlings, and house sparrows have adapted to living in urban environments and are harmful non-indigenous species like us. But there are many critters that are less charismatic and have had a more harmful impact to the organisms that have traditionally belonged in certain locales.

5. Habitat Destruction

One of the longstanding debates in the field of ecology was between Gleason and Clement (Worster, 1988), the argument being if associations and assemblages of organisms (ecosystems) were connected with each other because they had common connections (requirements) to their physical surroundings, i.e. climate and weather. Clement argued otherwise, that species were associated because of the interactions that had evolved between them. They had become connected or interconnected somewhat independently of the weather and environmental conditions. These organisms as a group, as a community of organisms, either visited with each other or decided to stay together, or they were circumspect among each other and did not visit or interact very often. A visit or interaction could be predation, symbiosis, or commensalism. Symbiosis and commensalism are terms that designate relationships that are mutually beneficial to all the organisms involved.

We have not formed symbiotic relationships with the natural environment. It is an old story—our encroachment and destruction of the natural environment. In the past we have altered the environment to serve our needs. Underneath the parking lot, underneath the city, there was once a field, wetland, stream, or forest which supported a variety of life. We were only recently reconnected to that world. Instead of that farm field there was once a native ecosystem or community of living interacting organisms. We now miss what was lost and have created preserves to protect pieces of what is left.

Human beings have been so destructive to the natural world that they have needed to be separated from the rest of the natural world in some definitions of nature. We have abilities, such as technology and the ability to purposefully alter our environment on a large scale. There is a general definition of nature synonymous with everything (also including the farm, city and cosmos), but there is also strict definition of nature that separates us from the rest of world because we are different. Our technological society is something the Earth has never seen the likes of before. We have not added diversity to the natural world; instead we have destroyed habitat and populations. There are areas in the world that are unpopulated, but we would be wise to reduce our population growth rather than increase it in the interest of maintaining a better standard of living for ourselves and our brethren around the world. Though we share abilities with other creatures of the planet, we are different than the other creatures that live on the planet. We are the only creatures that can be a steward of the globe, but we have left much destruction in our wake. We need to preserve pristine nature in order for other creatures to survive.

6. Overpopulation

It is not necessarily true that human beings, so far, have exceeded their carrying capacity on the whole planet, but that is clearly the case in many locations. The human population explosion has depleted the resources in many locations, and threatens the stability of human society in the long run (Ehrlich and Ehrlich, 2004). Millions of people are going to need to be relocated due to global warming, and overpopulation will continue to lead to famine and war. When resources become scarce due to overpopulation we will need to fight for them.

As a society we live, we reside on this planet, but it is more telling to say we also "ecologize" the planet in our own way. Other creatures merely follow the dictates of ecology, while we can choose our future. Acknowledging this also reminds us of our connection to and responsibility for the rest of life. We can do better by our children if we have less of them. An understanding of ecological principles derived from natural sciences reminds us of our fragility, and the fragility of other creatures that share the planet. As a society we are responsible for the other organisms on the planet as well. We have the capacity to acknowledge our situation and our effect on the rest of the planet and alter our behavior. Ecology has taught us we are part of an incredible interactive host of living things. Ecology has also taught us that we need to change in order to safeguard the other special organisms and connections on the planet. There are many ways to get involved. One is to change the way we live. Another is to lobby or vote for concerned parties. There are also many non-profit organizations that are concerned with the environment that could use your support. Your help and understanding is needed.

References

Bramwell, Anna (1989). *Ecology in the 20th Century*. New Haven: Yale University Press.

Callenbach, Ernest (1998). *Ecology: A Pocket Guide*. California: University of California Press.

Devall, Bill and Sessions, George (1985). *Deep Ecology, Living as if Nature Mattered*. Utah: Peregrine.

Ehrlich, Paul R. and Ehrlich, Anne H. (2004). *One with Nineveh: Politics, Consumption, and the Human Future*. Washington D.C.: Island Press.

Futuyma, Douglas J. (1984). *Evolutionary Biology*. Sunderland, MA: Sinauer Associates.

Merchant, Carolyn (1983). *The Death of Nature: Women, Ecology and the Scientific Revolution*. New York: Harper Collins.

Ricklefs, Robert E. (1983). *The Economy of Nature*. New York: Chiron Press.

Slobodkin, Lawrence B. (2003). *A Citizen's Guide to Ecology*. New York: Oxford.

Smith, Robert Leo and Smith, Thomas M. (2003). *Elements of Ecology*. San Francisco: Benjamin Cummings.

Strong, Douglas H. (1988). *Dreamers & Defenders: American Conservationists*. Lincoln, Nebraska: University of Nebraska Press.

Wilson, Edward O. (2002). *The Future of Life*. New York: Vintage.

Wilson, Edward O. (1984). *Biophilia*. Cambridge: Harvard University Press.

Pursuing Development and Protecting the Environment: Dilemma of the Developing World

4

Natalie Persadie, Rajendra Ramlogan

Embracing the notion of sustainable development is easier said than done, particularly in developing economies. There is now increasing recognition that the battle to preserve the global village from environmental harm cannot be won without the support of the developing world. Yet, developing countries are confronting challenges rooted in poverty and despair and tend to sacrifice environmental ideals in favor of economic imperatives. This paper paints a grim picture of the North-South divide and provides a statistical viewpoint of the challenge facing proponents of the principle of sustainable development. A case analysis of Trinidad and Tobago identifies the illusory effort by successive governments to promote environmental well-being. Findings show that governmental investments tend to privilege development issues (trade, energy, infrastructure) while investment in the environment is derisory and natural resources are seen as either a barrier to development or the equivalent of money to be cashed with obscene haste.

Introduction

The majority of developing countries are overwhelmed by poverty and the need for development. An analysis of key statistics pertaining to poverty and development reveals a somber picture of human misery in the developing world. It is, therefore, not surprising that much of the 20th century was spent in pursuit of developmental objectives primarily focused on economic wealth as the foundation of poverty eradication. It is generally accepted that all countries aspire to ameliorate the quality of life of their populations by improving living standards through economic growth prospects and development. The current dilemma deals with the desire of developing countries to attain a certain level of development using the status and model of developed countries as a standard, where development includes objectives such as the generation of wealth, the ultimate goal of which is the improvement in the quality of life. These countries seek to develop not only for the purposes of status, but for the more fundamental reason of the eradication of social ills. Developing countries are plagued with social inequalities and their defining features include poverty, inadequate social amenities, high rates of unemployment and a lack of proper infrastructure, among others. The chosen paths of development, though, are not always suitable to developing countries since they base their development policy on the paradigm set by the market economy.

Equally, leading theories on development show consistency with the aspirations of nations to achieve economic well-being and, ultimately, social upliftment. Not surprisingly, therefore, the pursuit of developmental objectives by developing countries usually comes into conflict with the ideals of environmental protection.[1] The problem that exists is that the issue of the environment is a low priority one in developing countries as they have other "problems" to look after. On the one hand there is the need for environmental protection as advocated by the developed world and, on the other, there is the fundamental need for development, which is also prescribed by the developed world. Developing countries have generally felt compelled to follow patterns of development determined by the developed world and, in so doing, they have used the rate of growth in GNP as a yardstick by which to measure their progress (The South Commission, 1990). The relationship that exists between development issues and environmental protection is such that one tends to be achieved at the expense of the other and it is usually the environment that becomes the hapless victim.

It should be noted that the state has a primary role in eradicating social inequities. In the developed world, civil society is very much involved in the decision-making process. Matters are not quite so organized in the developing world. There are, nevertheless, cases of civil society having influenced government regulations (Pargal and Mani, 2000). The fundamental problem that exists here is that the state is ultimately responsible for both allocating financial resources and ensuring environmental protection. The reasons for this may be threefold. Most environmental problems occur in public spaces, such as water or air, and it is therefore somewhat difficult to determine culpability. The state, as the custodian of public places and resources, is responsible for maintaining the integrity of public resources. Secondly, environmental problems are often transnational in nature and, as a result, fall under international law, which is the law between states and other actors on the international plane. This therefore implies that the obligation of protecting the environment lies at the level of the state. Finally, environmental degradation is largely the result of economic activity. It may be said that the private sector is generally concerned with capital gain and has no real incentive to protect the environment as this would cut into profit margins. Consequently, it falls to the state to protect the environment either by assuming direct responsibility or by instituting regulatory mechanisms to compel private sector contributions and 'green' corporate behavior.

The eradication of social ills and the protection of the environment, then, both depend on the same limited source for funding. This inevitably leads to prioritization and, hence, conflict. The environmental aspect obviously ranks lower on the list of priorities since the more obvious social problems such as health, poverty and education seem more urgent because the manifestations—starvation, sickness, unemployment, crime and vagrancy—are so flagrant. This would probably explain why the priorities of many developing countries do not necessarily include environmental protection.

Poverty and Development

Poverty

There exists an inextricable link between poverty, development and the environment. With poverty and inequality being defining characteristics of developing countries, the inherent need for development is manifestly clear if the plight of the poor is to be addressed. The problem that crops up, however, is that, in the developmental process, there is a tendency to wreak wanton havoc on the environment. It will be shown that the developmental drive generally forces environmental issues to a position of low precedence as it is said environment which provides the natural resources for development.

The concept of poverty may differ from culture to culture; however, with the relatively rapid Westernization of the world's cultures, the notion of 'lack' or 'deficiency' is common to most perceptions of poverty. Poverty, in its most extreme form, is a lack of basic human needs, such as adequate and nutritious food, clothing, housing, clean water, and health services. It may be defined as the condition that is said to exist when people lack the means to satisfy their basic needs. In this context, the identification of poor people, then, first requires a determination of what constitutes basic needs. These may be narrowly defined as those necessary for survival or as broadly as those reflecting the prevailing standard of living of the developed world. The first criterion would include people suffering borderline starvation, while the second would cover those people whose basic needs, as outlined above, while adequate to sustain life, do not measure up to that of the developed world.

One of the basic indicators of poverty is food and the need for an adequate diet. Poverty has been recognized as one of the major causes of hunger and malnutrition (Kent, 1984). Nutrients are needed to build tissues, produce energy and keep healthy (King and Burgess, 1993). The position of the developing world with respect to one area of nutrients will be examined. In the human body, food is the unique source of energy and may be expressed as caloric intake, as energy is measured in terms of calories (Bursztein, 1989). The importance of energy cannot be understated especially with regard to the performance of physical activities. The amount of energy or calories people need to obtain from food in order to keep healthy and active varies. One study, performed by King, has estimated that the average active man between the ages of 18 and 60 and weighing around 65 kilograms needs about 2,944 calories per day (1993). The developing countries received an average of 2,414.7 calories per day and the least developed countries 2087.5 calories per day (UNDP, 1996).[2] In relation to the study by King, this translates into developing countries receiving around 82% of the calories needed per day, and for the least developed countries about 70.1%. Extensive figures are not available for the developed countries but on available statistics, it would appear that people in these countries receive close to 103% of the calories needed per day (UNDP, 1996). With respect to the developing world, these figures are even more worrying especially as the statistics on per capita supply of calories do not reflect actual distribution. In some instances, on a general level, there may be enough food in a country but it is concentrated mainly in the top (richest) 20% of the population.

The consequence of food insecurity is often malnourishment, which primarily affects the destitute who are unable to produce or obtain required levels of nutrition. To make the situation worse, these people tend to inhabit unsanitary environments without access to basic services and clean water. Malnutrition also reduces the earning capacities and resources of households, so perpetuating their social and economic dependency, and reducing the effectiveness of governmental action to achieve developmental objectives (FAO, 1993). It was estimated in 1993 that 20% of the people in the developing world or approximately 848 million people were chronically undernourished (FAO, 1993).[3] The 1996 UNDP *Human Development Report* indicated that nearly 800 million people do not obtain sufficient food with about 500 million people being chronically undernourished (1996). Further, according to the 2001 UNDP *Human Development Report*, 17.6% of the people from developing countries are malnourished with the figure being 38% for the least developed countries. The figure for undernourished people in the high-income countries is generally not available with statistics obtained from only two countries with an average of 3.5%. Concerning children underweight for age based on percentage under age 5, the average for the developing countries is 23.6% and the least developed countries is 41%. Statistics are available for four high-income countries and the average is 6.75% (UNDP, 2001).[4]

Health is another basic indicator of poverty. Life expectancy is one of the primary indicators of health standards. It is true that today most people in the developing world are living longer than their ancestors. However, according to the 2001 UNDP *Human Development Report* the life expectancy of people in the developing world was 64.1 years in contrast with the figure of 77.8 for the developed world. The figure is even more dismal for the least developed countries where the life expectancy figure was around 51.3 years (UNDP, 2001). An example of an area of concern is communicable diseases. Although great progress was made in the 1960s and 1970s in combating communicable diseases, such as malaria, tuberculosis and cholera, these have returned in the late 1980s and 1990s with renewed virulence. Tuberculosis and cholera have struck with a vengeance in Latin America, with at least 1300 people dying in 1991 of cholera in Peru (Bello, 1994). Needless to say, the poor in the developing world have had to bear the brunt of the attack.

Quite often the weight of declining or poor health standards is felt first by the most vulnerable element of human society, namely children. The impact of poor health standards on children can also be discerned in the statistics dealing with the under-five mortality rate. Statistics from the 2001 UNDP *Human Development Report* reveal that the under-five mortality rate and the infant mortality rate per 1,000 live births are around 89 and 61 respectively for the developing world, with the corresponding figures being 159 and 100 for the least developed countries. The sheer magnitude of loss of infant life in the developing world is quite shocking when compared to the figures for infant mortality rate and under-five mortality rate per 1,000 live births in the developed world, which stands at around 6 and 6 respectively (UNDP, 2001).

Pregnant women constitute the other major vulnerable element of human society. The impact of low health standards on pregnant women is reflected in the low birth weight of babies (Dasgupta, 1993). According to the 2001 UNDP *Human Development Report*, an average of 7.6% of children born in the developing world had a low birth weight with 11.2% for the least developed countries. The corresponding figure for the developed world was 4.0% (UNDP, 2001). The implications of poor health are exacerbated by the lack of basic services, such as access to safe water and sanitation facilities. Statistics from the 2001 UNDP *Human Development Report* show that 70.1% of the people in the developing world had access to safe water and 64.9% access to sanitation facilities, with the figures worsening in the least developed countries to 57.2 and 47.2% respectively. Although data for the developed world are limited, current statistics show access to safe water and sanitation being in the vicinity of 99.7 and 99.4% respectively (UNDP, 2001). The lack of access to basic water and sanitation services fundamentally compromises the health of affected people.

The poor state of health care in the developing world can also be ascertained in the statistics dealing with health services. Between the period 1990 and 1995, 69.7% of the people in the developing world had access to health services, with the figure dropping to 54.1% for the least developed countries. Again the picture is different in the developed world with 99.5% of the people having access to health services (UNDP, 1996). A similar pattern is revealed when the data for physician per 100,000 people is examined. The 2001 UNDP *Human Development Report* shows that there was an average of 142.6 physicians per 100,000 people in the developing world and 66.1 physicians per 100,000 people in the least developed countries. The corresponding figure for the developed world was 251.8 (UNDP, 2001). Finally, the abysmal health record of the developing world is reflected in the statistics for the probability at birth of not surviving to age 40, the average of which, according to the 2001 UNDP *Human Development Report*, is 12.5% for developing countries and 33% for the least developed ones. In stark contrast, the corresponding figure for the developed world is a minimal 4.4% (UNDP, 2001). The probability at birth of not surviving to age 60 in selected developing countries is 19.9%, while it is 23.3% in selected least developed countries. In the developed world, the number stands at 10.3% (UNDP, 2001).

The need for adequate housing is fundamental as shelter is one of the most urgent necessities for human beings. The UN has recognized the tremendous problem of housing the poor and in 1976 hosted the UN Conference on Human Settlements (UNCHS) with the aim of addressing the issue. This led to the designation by the United Nations General Assembly (UNGA) of the year 1987 as the International Year of Shelter for the Homeless and the issuance of The Global Strategy for Shelter to the Year 2000. This document aimed at shelter for all by the year 2000 (Tolba and El-Kholy, 1993). Homelessness is a two dimensional concept. First, there are those that have no shelter, for example, those who live on the streets or in parks, and second, those whose shelter is inadequate, that is, improperly constructed, overcrowded and lacking in basic services, such as water and sanitation (Chege, 1995). Statistics on housing problems are not readily available on a global level. A press release in May 2002 from the United Nations Conference on Human Settlements states that there are some 100 million homeless people throughout the world, with women and children comprising the main victims. At least 600 million people, again mostly women and children, suffer from inadequate shelter (UNCHS, 2002). One study in India claims that there are about 700,000 pavement dwellers in Bombay and Calcutta, with over 5.5 million slum dwellers in Bombay alone (Bhaskara, 1989).

UNCHS has conducted the only major global housing survey and this was done between 1991 and 1992 among 35 cities in the developing world and 17 in the developed. This study found that an average of 1.0175 persons per 1,000 were homeless in low-income to mid-high-income countries, with the figure for low-income countries being 3.50. The corresponding figure for high-income countries was 0.97 (UNCHS and WB, 1993). In terms of the quality of housing, an analysis was performed on the percentage of houses built with permanent materials and not with temporary materials, such as straw, cardboard and cloth. The study revealed that an average of 13.5% of the housing in the low-income to mid-high-income countries were built with temporary materials, with the percentage figure rising to 33% for low-income countries. It was found that in developed countries, for all intents and purposes, 100% of the houses were constructed from permanent materials (UNCHS and WB, 1993). In terms of unauthorized housing, which includes both squatting and houses constructed without required building, land use or land subdivision permits, the average percentage of such constructions was around 30% for the low-income to mid-high-income countries, with a high of 64% for low-income countries (UNCHS and WB, 1993). The percentage of squatting alone was around an average 9.75% for low-income to mid-high-income countries, with around 17% for low-income countries (UNCHS and WB, 1993). The level of unauthorized buildings or squatting for the developed world was, for statistical purposes, insignificant (UNCHS and WB, 1993). While 2001 figures are admittedly scant for all categories, figures for squatting suggest an average of 4.7% for developing countries and 9.8% for least developed countries. In developed countries, the percentage was 3.5 (UNCHS, 1999a, 1999b). While developed countries also grapple with the problems of homelessness and inadequate housing, it would appear that the real problem lies in the developing world, especially in the rapidly growing cities.

Education is another important area that is being compromised by poverty. Statistics for 2001 suggest that the adult literacy rate for the developing world stood at 74.8% with the figure for the least developed countries being 51.9%. While statistics were limited, the average adult literacy rate for developed countries was around 92% (UNDP, 2001). The dismal adult literacy rate is supported by the equally disturbing trend in mean years of schooling. At 1992, the average years of schooling for the over-25-year-old segment of the population of the developing world was in the vicinity of 3.4 years and 1.5 for the least developed countries. The corresponding figure for mean years of schooling in the developed world was 9.67 years

(UNDP, 1994). These alarming statistics on adult literacy and average years of schooling suggest that much needs to be done to eradicate illiteracy in the developing world.

Equally important in the struggle for adequate educational standards in the developing world is the status of primary education. 1996 figures show that 66.3% of the first-grade enrollment completed their primary education with the figure dropping to 55.8% for the least developed countries. In developed countries, the percentage of children completing primary level education stands at around 96.3% (UNDP, 1994). The situation appears more calamitous when one considers that, for the same period, only 38.9% of the primary school entrants proceeded to secondary school with the figure being 18.95% for the least developed countries. The developed countries managed to ensure that 82.7% of primary school entrants proceeded to secondary schooling (UNDP, 1994). Statistics for 2001 for net primary and secondary school enrollment for developing countries stand at 54.6% and 29.9% respectively, while in the least developed countries the figures are a scant 29.1% and 7.3% respectively. For the developed world, the percentages were 73.8% and 59.6% respectively (UNDP, 2001). One positive improvement has been in the area of the number of pupils to teachers. For 1996 statistics show that this was at 33.6 and 21.6 for primary and secondary schools respectively for all developing countries. The corresponding figures for the least developed countries were 43.1 and 25 respectively. The industrial nations saw the number of pupils to teachers in the primary and secondary school institutions reach a satisfactory 18 and 14 respectively (UNDP, 1996). Notwithstanding the improvements in the developing world in the pupil-teacher ratios, there is still a long way to go, especially in the least developed countries, as the ideal ratio for primary school is probably in the vicinity of 23:1. Although large class sizes of an average of 45 is not the sole factor affecting school achievement, it increases the teaching load and reduces a student's access to the teacher. Very often, also, large classes are over-crowded and lacking in adequate furniture, a fact counter-productive to effective learning (Colclough, 1993).

Income levels have always been seen as important in the understanding of levels of poverty. Real income is often perceived as a measure of the command a person has over marketed goods and services, for example, food, clothing and shelter; although, it is now recognized that the income test, while useful, is a crude indicator of poverty. Real income does not indicate what goods are on offer or how much of each good on offer a person can and does procure. Nonetheless, changes in income levels generally point to movements in welfare levels (Dasgupta, 1993). The *World Development Report* 2000/2001 puts the average GNP per capita income for the developing world at US$ 2053.67, with the GNP per capita income for the least developed countries at around US$ 420.76. The state of income levels must be compared to that of the developed world where the GNP per capita stood at US$ 24,084.74 (World Bank, 2001b). Yet, the decade of the 1980s saw spectacular growth among nearly 1.5 billion people mainly in East Asia, where per capita income growth exceeded 7% per annum. In contrast, however, around 1 billion people among some 70 countries, mainly the least developed countries, experienced a decline in per capita income for the same decade (UNDP, 1996). Indeed, the proportion of the world's people blessed with per capita income growth declined from 54% to 37% between 1965-1980 and 1980-1993 (UNDP, 1996). The problems of income distribution are also well-reflected in international income distribution. The 2001 *Human Development Report* shows that the developed world accounted for US$ 23.9 trillion of the US$ 30.4 trillion global GDP (2001), with the share of global income of the richest 20% of the world's people rising from 70% in 1960 to 85% in 1991 (1996).

In assessing income levels to illustrate the wide disparity in income share, it is necessary to review income share within the framework of how much the richest 20% to the poorest 20% of population receive respectively. The 2001 UNDP *Human Development Report* shows that the wide disparity in income share between the top 20% and lowest 20% of the population in the developing world contributes to the brutal poverty that is crippling developing countries. For the figures available for the developing world, the poorest 20% of the people get an average of 5.6% of total income, with the figure declining even more tragically in Latin America to 3% (UNDP, 2001). The highest 20% receive an astonishing 45% of all income (UNDP, 2001). In Brazil, 40% of the households in that country received only a paltry 8% of the total national income, while the highest 20% received an astronomical 63.8% of the national income (World Bank, 2001b). When broken down into deciles, these figures seem more exaggerated: the poorest 10% of developing countries receive a miserly 1.9% of all income while the richest 10% receive as much as 27% of all income (UNDP, 2001). Income share distribution for the developed world is actually not any more equitable than the situation that obtains in the developing world. The poorest 20% of the population receives some 6% of all income while the richest 20% of the population receive 33.3% (UNDP, 2001). In terms of deciles, the percentages are 2.6 and 33% for the poorest and richest sections respectively (UNDP, 2001).

Income levels have also been used to determine the poverty line that has been defined as, "that income level below which a minimum nutritionally adequate diet plus essential non-food requirements are not affordable" (UNDP, 1994). Based on this poverty test, it has been estimated that there are some 1.3 billion people living in absolute poverty in the developing world. Similarly, based on the poverty line GNP per capita test of US$ 370 adopted by the World Bank, there are around 1 billion destitute worldwide. In some areas, the growth rate in number of the poor has been rapid. In Latin America, the number of people living in poverty rose from 130 million in 1980 to 180 million at the beginning of the 1990s (Bello, 1994).

Employment levels also play a part in the dilemma of poverty and it has been asserted that unemployment is one of the prime causes of poverty. It should be noted that while developed countries have social welfare programs to take care of their unemployed this facility is not available in developing countries. Employment provides people with a source of income, contributes to the output of goods and services and provides workers with a sense of dignity as they are provided with the means to participate in community activities (Griffin and Mckinley, 1994). The destitute are often unemployed, a condition that further undermines their ability to rise above the barriers of poverty. Thus, trends in employment are often indicative of the ability of a society to deal with poverty. In the developing world, 51% of the population forms the labor force with the figure dropping to 45% for the least developed countries. With respect to the developed countries, the labor force comprises around 50% of the total population (World Bank, 2001a).

Generally, figures on unemployment levels in the developing world are often unavailable or incomplete. However, for the few statistics available, an analysis for 1998 revealed a relatively high level of unemployment in the developing world. Bearing in mind that employment levels are generally understated due to poor data acquisition techniques and the absence of formal unemployment centers, the level of unemployment for 1998, the most recent figures available, was 8.6% for the developing world.[5] In the developed world, the unemployment level was 7% (World Bank, 2001a). Figures for the least developed countries were available for only 6 countries and from these the unemployment level is at 9.3%. While it would appear that the global village is generally experiencing similar levels of unemployment, two important differences must be noted. First, developing countries' unemployment levels are not as accurate as that of the developed world and unemployment percentages are not available for most of Africa, a region with major economic problems. Second, the social safety net present in most developed countries to assist the unemployed is generally absent in the developing world, partly accounting for the lack of enthusiasm in registering the unemployed, as no tangible benefits are available. Indeed, it has been argued that the concept of unemployment is not applicable to developing countries as, in the absence of a social net that allows people to spend time unemployed, people in the developing world must work regardless of how badly paid or how unproductive (UNDP, 1996).

It is clear from available evidence that poverty continues to stalk the developing world with devastating effects. Today, the issue of poverty is still as strong as in 1972 when the international community met for the first time to discuss poverty. At the United Nations Conference on Environment and Development (UNCED) in 1992, the second major international conference on the environment held 20 years after UNCHE, poverty once again assumed a dominant role. A joint report submitted by the Asian and Pacific Regions states "that the need for poverty alleviation is most urgent. The interaction between poverty and the environment sets off a downward spiral of ecological deterioration that threatens the physical security, well-being and health of many of the region's poorest people" (Economic and Social Commission, 1992).

Development

"The debate over the word 'development' is not merely a question of words. Whether one likes it or not, one can't make development different from what is has been. Development has been and still is the *Westernization of the world*" (Pieterse, 2000, p. 178).

Development is one of the most critical issues facing the developing world and has been described as "one of the most powerful of all Western ideas" (Freiburg and Hettne, 1985). Many authors have argued that the key to improving socio-economic conditions in the developing world is through development, where socio-economic issues include the ideals of environmental protection. This begs the question now as to the definition of the term development and the manner in which it is applied to these particular economies. While a holistic definition remains somewhat elusive, it is generally agreed that "the ultimate goal of a country's development process is an improvement in the well-being of its residents" (Yarbrough and Yarbrough, 1994) where both "improvement" and "well-being" are necessarily relative terms depending on the country's stage of development. For a developed country it may mean assuring its population a clean environment; while in a low-income country it may mean providing the inhabitants with the basic necessities such as food, shelter, clothing, education, and health and sanitation facilities.

Having regard to the challenges facing developing countries in pursuing developmental goals, it is necessary to assess the task that lies ahead especially as the common understanding that pervades developing countries is that the root cause of poverty is said to lie in the realm of economic backwardness or stagnation (Patel, 1991). For low-income countries, economic growth is imperative in the battle against poverty. This perception has influenced the manner of development of human society and over the last two decades the concept of development has converged with that of poverty. Both are being seen as essentially responsive to the other with developmental objectives being tailored to the attainment of anti-poverty objectives. Yet, developmental objectives of developing countries extend beyond the alleviation of poverty to include modernization goals, embracing areas such as communication and infrastructure.

With respect to communication, in 1998, there were 598 radios per 1,000 people in the developing world and 338 per 1,000 in the least developed countries in 1999. The corresponding figure for the developed nations was 813 (World Bank, 2001a).[6] Similarly, there were 155 televisions per 1,000 people in developing countries and 48 per 1,000 in the least developed ones. As expected, the frequency of television ownership was far greater among the developed nations, standing at 313 per 1,000 people (World Bank, 2001a). With respect to telephone linkages, there were approximately 150 telephone lines per 1,000 people in the developing world, with the corresponding figure in the least developed world being 23 per 1,000 people. The number of main telephone lines per 1,000 people in the developed world was 436 (World Bank, 2001a). On the level of infrastructure, one example of a developmental objective is roads. The percentage of roads in good condition (paved) amounts to 54.6% of the roads in the developing world, with the figure for the least developed countries being a close 51%. In the developed world, some 86.5% of the roads are similarly classified (World Bank, 2001a).[7]

Poverty, Development and the Environment

Having seen the justifiable preoccupation with poverty and development in the developing world, there is a need to understand the undeniable link with the environment. Poverty and development have contributed to environmental degradation in several ways. For the poorest in the world, the environment is the means to survival, that is, the forests, rivers and land are relied upon to provide basic needs. For example, trees provide food crops, timber for building, fuel for cooking, fodder for animals and fibers for weaving (Davidson, 1992). The World Bank has asserted that the poor are both victims and perpetuators of environmental abuse. Lacking in resources and technology, impoverished farmers often engage in faulty agriculture practices such as 'slash and burn' agriculture, cultivation of erosion-prone hillsides or movement into tropical forests where crop yields often decline quickly after a few years. The needs of poor families are often short-term and this prompts the fast utilization of natural resources, such as excessive cutting of trees and failure to replace soil nutrients (World Bank, 1992).

The rapid deterioration of the environment has made the daily struggle to maintain one's existence even more monumental. This has led to the harvest of nature's bounties beyond sustainable levels in a manner deleterious to the continuing integrity of the environment. The result has been environmental problems such as massive deforestation and growing desertification in some areas like the Sahel region (Smith, 1994). The result is often the migration of poor families into forested areas in search of new lands. It has been estimated that 13% of the people in the developing world are landless and almost 60% of the households lack adequate land for subsistence (Davidson, 1992). Constantly pushed by economic forces, these landless people quite often move into marginal environments that tend to be more ecologically fragile, such as arid lands, tropical forest, savannahs, steep mountain slopes and mangroves (Gupta, 1988).

Davidson has asserted that 60% of the world's poorest citizens live in ecologically vulnerable areas with the pressures of settlement and cultivation in these areas leading to further and more rapid environmental degradation (1992). More so, developing countries appear to suffer more at the hands of so-called "natural" disasters. Inevitably, this occurrence seems more prevalent in the poorer developing nations like Bangladesh.

It would appear that the world's poor, by intensifying their use of natural resources by improper environmental practices such as deforestation, significantly contribute to the intensification of the impact of natural disasters (Pearce and Warford, 1993). The result of their actions ensures that the damage caused by natural occurrences is rendered much greater than what would have prevailed had the environment not been abused.

Developmental projects have also been undertaken without regard to the environmental ramifications. The frequent failure to assess the impact of projects like dam construction, road building and railways has added to the many environmental disasters unfolding in the developing world. One example of such a disaster is the Projeto Grande Carajas in north-east Brazil. It occupies 900,000 square kilometers; more than 10% of the land area of the state of Para, whose topography and ecology are being irreversibly altered by the project. In 1986, 890 kilometers of railroad track were laid for Carajas' sole use, and the project currently sends three 160-car trains per day, each one carrying 15,000 tons of iron ore, to the Atlantic port of Sao Luis. The ironworks are powered by charcoal, for which 200,000 hectares of virgin forest are cut down each year; eucalyptus trees are mainly used for "reforestation" because they grow rapidly. Rio Tocantins, one of the largest rivers in the Amazon basin, has been dammed at Tucurui for hydroelectric power, and at the same time 2,400 square kilometers of forest have been submerged. Waterways are already being clogged by soil from cleared land areas causing flooding. The failure to clear the land before flooding means that dams are polluted with rotting vegetation enabling the breeding of disease-carrying insects. The makeshift towns that have sprung up along the development have added to the environmental catastrophe that is the Projeto Grande Carajas (Davidson, 1992). The 23,000 workers in the region, some of whom have lost their jobs and have settled there permanently, have come into the usual problems experienced by other Brazilian "pioneer" areas in the past: slash-and-burn, over-settlement, garbage, prostitution, crime and poverty (Carajas, 1998).

Very often, economic growth is based on the unsustainable exploitation of natural resources and pollution of the environment. Indonesia is frequently cited as a nation that has achieved economic growth together with human development. However, in the 1980s, it had an annual deforestation rate of 1% leading to a loss of 1.2 million hectares of forest per year. Similarly, Thailand, with its spectacular economic growth in the 1980s, saw its forest cover reduced from 55% to 28% between 1961 and 1988. Nature appears to be paying the price for the economic rejuvenation of many societies. Pollution has risen in many developing countries with high economic growth, as the quality of the environment is ignored in favor of economic successes. In Taiwan, less than 1% of human waste is treated and this has led to one of the highest rates of hepatitis-B in the world. In South Korea, much of the tap water available for drinking purposes is contaminated, and the capital Seoul is further rated as one of the five worst cities in the world with regard to air pollution (UNDP, 1996).

Finally, and perhaps most importantly, poverty and development affect the ability of developing countries to pursue environmental objectives by diverting national income. Government must take measures to combat poverty and hasten development. As the lead spender with regard to environmental objectives, the environment quite often slips down the ladder of government fiscal priorities. The price for the failure to address poverty and development can be a high one, as poverty has the ability to adversely affect peace and stability. Poverty is capable of breeding social and political conflicts and undermining fundamental human values (Jazairy, 1992). Thus, there is emerging recognition of the dynamic interactions of poverty, development, stability and the environment. Yet, nations are expected to pursue economic development in order to provide basic needs and achieve modernization goals. The pursuit of these objectives is paramount and until developing countries can sufficiently reduce poverty to manageable proportions, environmental goals will remain peripheral and concepts such as sustainable development will remain as high sounding rhetoric bereft of substance (UNCED, 1992). In echoing the words of

Indira Ghandi, Robert Mugabe, the former President of Zimbabwe, stated that, "preventable poverty is one of the major causes of environmental degradation today. Poverty pollutes our environment. Those who are poor and hungry will often destroy their immediate environment in order to survive. Their livestock will overgraze the grasslands and in growing numbers will crowd congested cities. They will overuse marginal lands. This explains why the greatest environmental change is occurring in developing countries. These countries are poor" (1991, p. 11).

Development and the Environment: A Case Analysis of Trinidad and Tobago

The state is the critical player in ensuring effective environmental management. The fact that the state is the primary financier of environmental investment makes it imperative that the state embraces the desire to deal with environmental issues at both the policy and implementation level. It has been argued that developing countries have not yet attained that stage of development where they can afford to give priority to environmental management. It is true that developing states sometimes attempt to introduce measures to deal with the problem of environmental degradation; however, the problem is that rapid population growth generally unhinges the efforts of the state. Developing countries would only be able to pursue environmental protection policies when they have achieved certain developmental objectives and can afford to divert some of their financial resources for 'lower' priority issues.

Currently, there is still an inherent need to develop the economic and social infrastructural bases of these countries, as it has been persuasively argued that the application of appropriate developmental policies is the only way to eliminate social inequalities. The intrinsic problem that exists is that it is the state that is ultimately responsible for allocating the finite financial resources available to it. This is not to say that civil society, in the form of community groups, does not help in the decision-making process. While it would appear to be less organized than the groups mobilized in the developed world, this type of collective action can supplement formal governmental regulations (Pargal and Mani, 2000). At the end of the day, this entity is responsible for the economic and social development of the country; it is, however, also responsible for the protection of the environment. This is where the inherent conflict lies, as the state must now decide which is more important and allocate the resources accordingly.

The fact of financial prioritization shall comprise the general trend of thought in the following section. Investments into an institutional framework for environmental management will be examined for one local environmental agency and the environment generally in Trinidad and Tobago, which will certainly be compared to the funding budgeted for trade and investment to clearly illustrate financial prioritization.

The Statistics: Government Allocation Compared—Environment versus Trade and Development

It is obvious that, as a developing country, government expenditure is geared toward facilitating developmental objectives and is not necessarily as concerned with issues of an environmental nature. In the review of the *Public Sector Investment Programme* for the year 1998/1999, the Overview states,

> The Public Sector Investment Programme (PSIP) 1998/1999 focused continuing and advancing programmes and projects for the development and enhancement of the economic and social infrastructure [*sic*] base of the economy. The aim of the 1998/1999 PSIP was to increase as well as sustain the level of economic growth by strengthening the base for facilitating economic activity. To this end, resources were allocated to priority sectors identified as agriculture, roads and bridges, education, health, housing and settlements, water and sewerage, social and community services and public administration, as well as for strengthening the capacity to protect and conserve the natural environment (Ministry of Planning and Development, 1999, p. 1).

While the objectives of the PSIP are rather commendable, it is interesting to note just exactly when mention is made of the need for environmental protection and conservation. While it is true that housing and agricultural policies could be environmentally-friendly, they generally tend to concentrate on infrastructural development. Equally interesting is the fact that 1998/1999 is the first and only fiscal year in which environmental concerns are actually stated in the PSIP objectives. The PSIP 1999/2000 makes no explicit reference to environmental management in its objectives, but seeks to enhance "the social well-being of the people, having laid the economic fundamentals on a solid footing" (Ministry of Planning and Development, 1999, p. 26). The objectives for previous years emphasize the importance for developing economic and social infrastructure that could ultimately provide "a platform for sustained economic growth while improving the quality of life through the enhancement of social services" (Ministry of Planning and Development, 1996, p. 1).

This priority given to development and growth is reiterated in the Government of the Republic of Trinidad and Tobago's *Medium Term Policy Framework 1998-2000*. The medium term goals and objectives state:

> Government's main objectives over the medium term are to accelerate the growth and development of the economy; facilitate the diversification of the country's productive base; and ensure an improved standard of living for all. The broad policy goals over the medium term are identified as follows:
>
> - Promote strong economic growth;
> - Reduce unemployment;
> - Contain the inflation;
> - Ensure equity within the society; and
> - Protect and preserve the environment (Ministry of Finance, 1997, p. 4).

For the fiscal year 1999/2000, allocation for all sectors amounts to TT$ Mn 1567.5 (US$ 249).[8] Given past trends, total investment expenditure usually amounts to a percentage of planned investment. For 1998/1999 planned investment was TT$ Mn 1097.1 (US$ 174), but actual expenditure stood at TT$ Mn 667.5 (US$ 106), or 61% of planned investment

(Ministry of Planning and Development, 1999, p. 1). For 1997, expenditure amounted to 70.9% of planned expenditure (Ministry of Planning and Development, 1997, p. 1). Of the $ 1567.5

Table 1: Financing the Public Sector Investment Programme 1999/2000

Sources of Financing	$Mn	%
Central Government Programme	**1552.485**	**99.043**
External Sources	**479.397**	**30.584**
Loans	**435.044**	**27.754**
Inter-American Development Bank	282.753	18.039
International Bank for Reconstruction	103.641	6.612
Caribbean Development Bank	48.65	3.104
European Investment Bank	0.00	0.000
Grants	**44.353**	**2.830**
Inter-American Development Bank	3.928	0.251
International Bank for Reconstruction	3.090	0.197
European Community	35.095	2.239
Other External Financing Agencies	2.240	0.143
Domestic Sources	**1073.088**	**68.459**
Loans	**131.890**	**8.400**
Republic Finance and Merchant Bank	131.890	8.400
General Revenues	**916.198**	**58.450**
Government of Trinidad and Tobago	916.198	58.450
Road Funds	**25.000**	**1.595**
Government of Trinidad and Tobago	25.000	1.595
State Enterprises Programme	**15.000**	**0.957**
Domestic Sources	**15.000**	**0.957**
General Revenues	**15.000**	**0.957**
Government of Trinidad and Tobago	15.000	0.957
GRAND TOTAL	**1567.485**	**100.000**

Source: Ministry of Planning and Development, 1999, p. 28

Table 2: Public Sector Investment Program (PSIP) 1999/2000
Analysis of Sectoral Allocations

	TT$Mn	%
PRODUCTIVE SECTORS	**22.20**	**1.42**
Agriculture	22.20	1.42
ECONOMIC INFRASTRUCTURE	**531.90**	**33.93**
Agriculture, Fisheries, Forestry	76.42	4.88
Manufacturing	17.45	1.11
Drainage	26.65	1.70
Electricity	2.39	0.15
Environment	8.30	0.53
Land Acquisition	11.00	0.70
Roads and Bridges	260.15	16.60
Tourism	15.11	0.96
Transport and Communication	2.10	0.13
Water and Sewerage	13.75	0.88
Other Economic Services	93.60	5.97
Urban and Regional Development	5. 00	0.32
SOCIAL INFRASTRUCTURE	**716.34**	**45.70**
Education	235.20	15.00
Health	157.21	10.03
Housing	75.92	4.84
Social and Community Services	229.37	14.63
Human Resource Development	16.65	1.06
Training and Support for Employment	2.00	0.13
PUBLIC ADMINISTRATION	**268.89**	**17.15**
Administration	175.32	11.18
Public Order and Safety	93.57	5.97
PLANNING/PROJECT PROGRAMME DEVELOPMENT	**28.17**	**1.80**
Planning and Project Development	28.17	1.80
GRAND TOTAL	**1567.5**	**100.00**

Source: Ministry of Planning and Development, 1999, p. 80

million allocated for the PSIP for fiscal year 1999/2000, a total of 68.459% is derived from domestic sources.

While it is true that domestic sources provide 68.5% of all funding, there is still a relatively heavy dependence on funding from external sources. More importantly, is the actual sectoral allocation of planned investment.

One can immediately see that development is not solely focused on economic infrastructure, but even more so on social infrastructure. Nevertheless, within the breakdown for economic infrastructure there is obvious prioritization. One must take into account that allocation to environmental issues cannot be checked singly since allocation to all agencies with environmental responsibilities must also be included.[9] Even in this table, there is overlap. While allocation to the environment is separately considered, environment-related issues can also be found under agriculture, fisheries and forestry; manufacturing, which includes environmental impact assessment specifications and the development of a computerized environmental system; and planning and project development, which includes a natural resource and environmental management program. That aside, it is evident from the table that development is given great importance. The development of roads and bridges alone has been allocated 16.6% of all planned expenditure. Tourism and manufacturing account for 2.07% of all investment. The environment, however, has only been budgeted TT$ Mn 8.3 (US$ 1.3), or little over ½% of total investment. Compare this to the $ 76 (US$ 12) million estimated to support business expansion and industrial development in the sectors of agriculture, fisheries and forestry, which is a significant 4.8% of all expenditure.

For 1998/1999, more emphasis was placed on economic infrastructure with 52.15% allotted to economic development. An enormous 21.18% was granted for roads and bridges alone; 3.03% to manufacturing and tourism; and 1.38% only to the environment. Of the $ 6.7 (US$ 1) million allocated to the environment, namely the continued capacity development of the Environmental Management Authority ("EMA"), only $ 4.1 million (US$ 0.65) was expended. Resources totaling

$ 20.4 million (US$ 3.2) funded from general revenues were allocated to the Tourism and Industrial Development Company (TIDCO) of Trinidad and Tobago in 1998/1999 for facilitating and promoting the development of the industrial base of the economy (Ministry of Planning and Development, 1999).

TIDCO was incorporated in 1993 as the principal marketing, promotion and development agency in the non-oil, non-energy sector in Trinidad and Tobago. While the company commenced operations in mid-1994, (only half year before the institutionalization of the Environmental Management Authority), it has already assumed great importance. For operational purposes, the company has been subdivided into seven divisions: the Office of the President, the Division of Trade and Industry, the Division of Tourism, the Division of Government Policy and Special Projects, the Division of Finance and Administration, and the Property and Industrial Development Company of Trinidad and Tobago (PIDCOTT) (TIDCO, 1999).

One look at the Consolidated budget for the Company will illustrate the importance placed on development. As a whole, the company had a budget projection of TT$ 150,756,313.00 (US$ 23,929,574) (TIDCO, 1998) for the year 2000 and obvious emphasis was placed on the divisions dealing with industrial development, trade and tourism. These sections accounted for 58.5% of all of TIDCO's expenditure. The Government of this country is obviously concerned with developing the economic base of the country. Based on the allocations to the respective sectors, it affords little attention to environmental issues as its limited financial resources force it to prioritize.

What this data illustrates simply and clearly is that priority is definitely given to developmental objectives. The emphasis of developing countries is to develop their economic and social infrastructural bases. This generally means that the environment is not prioritized and is put on the back burner as resources are shared according to the importance afforded to the respective issues. The environment is clearly not treated as being very important.

Investment in an Institutional Framework for Environmental Management in Trinidad and Tobago among Enforcement Agencies

This section will look at the development of the Wildlife Division, an important agency with responsibility for protection of the environment as a microcosm of the problems facing enforcement agencies in Trinidad and Tobago. This will be used to highlight problems affecting its efficiency (in its capacity as an environmental-related agency), such as the resources (human, technical, financial and research) available to it and its enforcement capability. This agency is developed by internal funding and will therefore reflect financial prioritization as evidenced by the myriad inadequacies with which it must deal as a direct result of the poor financing they receive.

The Wildlife Section is part of the Forestry Division which falls under the Ministry of Public Utilities and the Environment.[10] The issue of wildlife protection first came under the Forests Act which dates as far back as 1915 and wildlife management constituted part of the functions of the Forestry Division. The Conservation of Wild Life Act was passed as Act No. 16 of 1958 and provided for the establishment of a Wildlife Conservation Committee. As the name of the Act implies, the purpose of the Wildlife Section (separate from the Committee) is to "discharge its responsibility for the research and management of the wildlife resources of Trinidad" (Ministry of Agriculture, 1999a).

The major constraints of the Wildlife Section are a lack of staff and funds. A Senior Ministry official complained about the quality of the staff; he stated that there was an inherent need for more trained staff. In addition, the Wildlife Section is always under-funded as the releases (money allocated by the government) are always less than the allocations (what the division requests). For 1997, out of a total of TT$ 3,877,745.13 (US$ 615,515) spent on the development program for the Forestry Division as a whole, only TT$ 150,000.00 (US$ 23,810) was spent on wildlife (Ministry of Agriculture, 1999b). For the 1999/2000 Fiscal year TT$ 76.42 million (US$ 12.13), or 4.88% of total budgetary allocation, was allocated to agriculture, fisheries and forestry

(Ministry of Planning and Development, 1999, p. 80). Of this, only TT$ 2.4 million (US$ 0.38) was allocated to the Forestry Division (Ministry of Planning and Development, 1999). It would be out of this amount then that the Wildlife Section would be allocated funding.

In 1998, there were nine vacancies for the position of Game Warden I and one vacancy for the position of Game Warden II. The Game Wardens are required to increase the capacity of the Wildlife Section. However, the Service Commission appointed only the Game Warden II (the Director of Forestry); the other nine positions remain unfilled. Eight years ago, Cabinet agreed to appoint 280 honorary Game Wardens to assist the regular Game Wardens. Although the appointments were made, the impact on the effectiveness of the Wildlife Section has been barely noticeable. This is compounded by the inability to supervise all the honorary Game Wardens properly.

There is a definite inadequacy with the enforcement capabilities of the Wildlife Section, which are also to be attributed to staffing constraints. Although Trinidad and Tobago is a signatory to the Convention on the International Trade in Endangered Species of Wild Flora and Fauna ("CITES"), which regulates the import and export of species on the negative list, it is often difficult to control the illegal trade of protected and endangered species. The Cedros Peninsula, at the southwestern tip of Trinidad (mere kilometers away from Venezuela), has been cited as a major point that facilitates this trade. While the police and defense force assists them, this is often not enough.

Another huge problem facing the Wildlife Section is illegal hunting. However, a major constraint to the Wildlife Section is the inadequate and unsuitable weapons available to the Game Wardens. The weapons are outdated and are no match for the poachers who are equipped with modern, sophisticated weapons. Therefore, even though the Wildlife Section can call on 280 honorary Game Wardens, their ability to act effectively is hindered by the lack of modern technology.[11] The Wildlife Section, which operates largely in remote localities, is particularly hampered by an inadequate fleet of vehicles. The efficiency of the Wildlife Section has been compromised and, as a result of the financial constraints, it is ineffective. The stark reality, therefore, is that the Wildlife Section lacks an appropriate enforcement capability and the legislation is breached with impunity.

Natural Resources and Development: Friend or Foe to the Environment?

The other dimension to state support of the environment lies in the polemic relationship with development and natural resources. While natural resources are certainly a significant platform for achieving development within economic parameters, it can suffer immensely during the monetization process. It can be cogently argued that the natural resource base of developing countries is perceived either as commodities to be harvested for the purposes of financial gain, or as obstacles to the development process.

Natural Resources: Commodities

The Economic Development of Trinidad and Tobago and the Energy Sector

Trinidad and Tobago provides a prime example of a natural resource-based developing economy. Its economic development is driven largely by the exploitation of its natural resources, as illustrated by its dependence on sugar in an earlier era, and then on oil and natural gas. However, in the country's quest for development, it is interesting to note its stance on environmental protection as concerns its largest foreign exchange earning sector—the energy sector.

In 1998, total domestic oil production was 44.7 million barrels at an average rate of 122.627 barrels of oil per day ("bopd"), a decline of less than one percent from 1997. However, average daily oil production rose to over 130.000 bopd towards the end of the year (MOEEI, 2000). As the statistics show, crude and refined petroleum accounted for as much as 88.6% of total exports in 1980 and generally accounted for more than 70% of all exports for the period. Current statistics gleaned over the years attest to the fact that the energy sector remains the largest single contributing sector to the country's economy as indicated in foreign earnings, gross domestic product and Government revenues. The following statistics illustrate the sector's continuing contribution to key economic indicators in 1997:

Gross Domestic Product:	$ 9.7 Billion or 26.6% of the total GDP of $ 36.4 Billion
Government Revenues:	$ 2.1 Billion or 21.2% of total Government Revenue of $ 9.9 Billion
Foreign Exchange Earnings:	$ 11.2 Billion or 71% of foreign exchange earnings of $ 15.9 Billion (MOEEI, 1998, pp. 7-8).

For 1998 "excluding the petrochemical industries, the energy sector was expected to contribute 23% to GDP and 76% to exports..." (MOEEI, 2000).

The peak in oil production in 1978 (of 12 million tons per year [MOEEI, 1994, p. 2]) encouraged exploration that resulted in a shift in the ratio of gas to oil reserves. In the face of declining oil reserves, these exploration activities revealed the existence of significant reserves of natural gas. Natural gas, which had been hitherto treated as a nuisance product and flared, began to assume an economic significance of its own (MOEEI, 1998). It is also used as the petrochemical feed stock in the production of ammonia, methanol and urea. It provides an energy source for non-energy products such as steel, iron carbide and other minor manufactured products (MOEEI, 1998).

A major development of the natural gas industry is the liquefaction of natural gas for export. This is projected to be the fastest growing sub-sector of the national hydrocarbon industry. By the year 2000, natural gas consumption was expected to reach 1400 million cubic feet per day ("MMcfd") from the then current level of 900 MMcfd. Some 450 MMcfd of this additional gas will be Liquefied Natural Gas ("LNG") exports, while additional supplies will be consumed as fuel for power generation, feed stock to new petrochemical plants, and fuel for additional metal furnace plants (MOEEI, 1998).

The Energy Sector and the Environment

Given the extreme importance of the hydrocarbon sector to the economy, it is now relevant to examine the attitude of the (energy) policy makers to environmental protection. While both pre- and post-independent budget speeches were significantly shaped by the obvious dependence on the then oil industry, there was little attempt to formulate a policy concerning the energy sector. Even despite the passing of the Petroleum Act of 1969, the first policy paper only appeared some twenty years afterwards. It was only with the change of government in 1986, when the National Alliance for Reconstruction ("NAR") came into power, that the policy paper was written. As the 1986 Energy Policy states,

> This Paper is a first step towards documenting a comprehensive Energy Policy for Trinidad and Tobago. It involves consolidation of previous and existing policies, updated and articulated into a more fully integrated energy policy document (Ministry of Energy, 1986, p. 4).

Given the political party's concerns of environmental issues, the policy includes a theoretical commitment to safeguard the environment. However, the policy states no actual prescriptive measures on sustainability or pollution prevention, nor does it set any deadlines for the implementation of any such measures. Nevertheless, it does acknowledge that "the disposal of wastes from energy will continue to degrade the environment..." (Ministry of Energy, 1986, p. 71). The policy vaguely states that its two main goals are to:

- Provide adequate energy for the country's needs...in the most efficient and economical way, while ensuring that...the long-term quality of the environment is maintained at an acceptable level; and
- Enable Trinidad and Tobago to fulfill its obligation to several UN organizations to which it has pledged to contribute to the solution of environmental problems... (Ministry of Energy, 1986).

An "acceptable level" needs to be clearly defined. There need to be specific measures to ensure efficiency. There is also need for precision in terms of the country's ability to fulfill its international environmental obligations. This first policy left much to be desired for environmental concerns.

The 1992 Energy Policy offers no more commitment than does the first. This particular policy states that,

> The Government will place further emphasis on environmental considerations and the promotion of the use of environmentally-friendly energy substitutes as alternative transportation fuels (MOEEI, 1992, p. 27).

Again, this is imprecise and offers no suggestions as to how to achieve such considerations. In detailing the parameters for formulating an energy policy, of eleven concerns expressed, environmental concerns were the tenth mentioned. Although it was not expressly stated that the measures appeared in order of priority, it is feared that the order in which the measures were presented reflected the importance attached to environmental policy. The situation remains largely unchanged for the 1994 Energy Policy.

The government in office in 2000 (the United National Congress) was allegedly also concerned with the phenomenon of environmental degradation and as such, endeavored to include environmentally-friendly practices in its policies. The most recent Energy Policy (1998) during their reign also makes sweeping statements. "The national drive for efficiency, effective management, environmental preservation, sustainable development and a total quality nation must be reflected in the country's energy policy" (MOEEI, 1998). This policy takes a slightly different outlook: it looks at environmentally cleaner fuels. Given the shift to natural gas utilization, it mentions the use of compressed natural gas (CNG) as a vehicular fuel. While this has been introduced, there are very few stations that dispense CNG, and the number of vehicles equipped to use CNG is also minimal. This policy makes mention of sustainable development, yet there are no practical measures in place to support this. Interesting to note also is that the section on environmental matters is always the last section of each policy. All the policies clearly illustrate the lack of commitment to environmental protection in the interest of development.

Equally fascinating is the declaration made by the then Minister of Energy himself. He claimed that "the Ministry's policy was zero tolerance for environmental damage caused by energy sector companies"; but in the same breath, he admitted that while "the sector was the major contributor to the nation's development and to the national economy, it was also the biggest contributor to environment degradation" (De Coteau, 2000).

The attitude towards the environment in the management of the energy sector reiterates the fact that as a developing one-commodity country economy, the government is not prepared to compromise the petroleum industry, given its significance to the economic development of the country, for the sake of the environment.

Natural Resources: Obstacles in the Way of Development

The destruction of the environment for the purposes of development is the sometimes necessary alternative to the actual direct utilization of natural resources for economic gain. This is particularly true of the developing world where "the sheer interests of survival outweigh environmental concerns in the minds of many" (Ramlogan, 1999). There are many often more pressing (social) issues to be dealt with, especially *la mis ère*. This disregard for the environment assumes an international character, especially where the construction of large dams and hydro-projects is concerned, with a view to improving agriculture and providing energy.

Governments of developing countries often ignore environmental issues for the purposes of development. The construction of the Tehri dam in the Garhwal region of Uttar Pradesh (India) readily illustrates the attitude of governments in the developing world. The Tehri dam project was conceived as far back as 1949 but due to numerous controversies surrounding its construction, the completion deadline was pushed to 1996. The projected cost of the Tehri Dam has gone up from

$ 612 million in 1994 to $ 1.2 billion in 1999. With a height of 260 meters (855 feet), the dam will be the fifth tallest in the world. Its reservoir will completely submerge Tehri Town and 40 villages, and partially submerge 72 villages. Thousands of families will be displaced with no apparent scheme for rehabilitation. According to the 2002 Status Report of the Public Works Department of Tehri, the Dam will displace 12,547 families. This estimate excludes a large number of people who will lose their lands but have not been officially recognized as project-affected. This dam has been opposed by a wide cross-section of the society, including high profile protestors such as Arundhati Roy. It is being built in an ecologically fragile area, which also happens to be located on an active geological fault line. Despite widespread protest, the Indian government has continued the construction process with a typical "business as usual attitude" (Hassan, 1991). Under construction since 1978, the final phase of reservoir filling was scheduled to start by December 2002. Power production was planned to begin in August 2003. The two lower tunnels of the dam were closed in December 2001 submerging the main bridge leading to Tehri town and nearby areas. The last two tunnels, which are at a higher level, were scheduled to be closed by December 2002 (IRN, 2002).

In Trinidad, on a much less exaggerated scale, there also exists the development/environment trade off, as illustrated by the controversial port development project at Invader's Bay, Port of Spain. The Port Authority of Trinidad and Tobago (PATT), the Project Manager, embarked in an historic undertaking to deepen the Port of Spain harbour, from its draft of 9.75 metres to 12 metres (Port Authority, 2000b) which would involve reclaiming much of a nearby mangrove. In order to facilitate the growth of the city "much of the marshland was claimed or reclaimed in the name of development" (Port Authority, 2000a). This particular project involves the reclamation of 28 hectares of land at Invader's Bay. "This land reclamation project is a critical component in government's overall plan for the development of the West Coast of Trinidad"

(Port Authority, 2000a), much to the disappointment of the local environmentalists.

In an Environmental Impact Assessment (EIA) for the proposed port expansion prepared in 1998, it was recommended that "the dumping of the dredge spoil [be] at approved sea sites...as it is more economic than onshore disposal" (Meredith, 2000b).[12] This has, however, since changed. "The dredge spoils from the channel deepening are now to be used to reclaim Invader's Bay instead of being dumped at sea as proposed originally" (Chackhan, 2000). One of the conditions of the EIA was the restoration of the mangrove. This will not be affected despite the fact that the *National Environmental Policy* (*NEP*) specifically discourages any form of development in wetland areas (Meredith, 2000a). The *NEP* states that the government will:

(a) pursue a policy in its development projects of no net loss of wetlands;
(b) institute mechanisms for the restoration of degraded wetlands, as far as possible, to their original state;
(c) discourage dredging, filling and other forms of development in wetland areas... (EMA, 1998, p. 11).

The port expansion was obviously in direct contravention of the *NEP*. While the then government claims to have followed all the correct procedures, it is manifestly clear that practice proves otherwise. Further aggravating the situation is that the project began without the approval from the relevant authorities at that time—the Town and Country Planning Division. To some, this project is an apt example of the Government's blatant disregard for the environment. As Mark Meredith aptly put it, "Money and master plans clash with a mangrove in the city. Is this the cost of economic growth?" (2000a). Fishermen and Friends of the Sea, a local environmental non-governmental organization (NGO), fought the project through an aggressive campaign in the vicinity of the development to attract public sympathy and to force a reversal of the decision to develop the port. Although the NGO was able to increase public awareness about the issues at hand, unfortunately, not much success was achieved in halting the project.

An interesting point that can be made was a recent incident in 2002 where a Canadian oil company, Talisman (Trinidad) Petroleum Limited, sought to explore for oil in the Nariva Swamp, a Ramsar listed site. What was interesting was that the local environmental enforcement agency sought to prohibit any oil exploration work in the Nariva Swamp while Talisman argued for wise use and sustainable development. The Environmental Commission of Trinidad and Tobago, in delivering its first judgment on December 16th, 2002, dismissed the notion of prohibition and indicated that the local environmental agency had to assess the application and allow Talisman to make its proposal for wise use.[13] This decision, in our opinion, however, did not reflect a policy shift to primacy of environmental policy but local politics. From the date on which the appeal from the decision of the local environmental agency was filed by Talisman in the Environmental Commission, the company faced a barrage of public criticism, including comments attributed to senior officials of the local environmental agency. One member of the Board of Directors was cited in a newspaper article as a party submitting comments on the application by Talisman to explore for oil in the Nariva Swamp on behalf of a local environmental NGO (Meredith, 2002). Therefore, much cannot be read into the actions of the local environmental agency as signaling that development based primarily on economic considerations is being transformed in favor of environmental sustainability.

A recent issue arose in Trinidad and Tobago with respect to the production of liquefied natural gas. A multinational consortium, known as Atlantic LNG Company of Trinidad and Tobago, whose shareholders include British Petroleum and British Gas, established three facilities known as "trains" for the production of liquefied natural gas between 1995 and 2002. The consortium proposed doubling the capacity of the first three trains in a single train (known as Train IV). Many locals living in the vicinity of the existing trains have opposed Train IV on the basis of allegations of environmental disturbance caused by factors such as air, noise, dust, light and vibration and resulting health concerns. These locals have called for the establishment of a buffer zone and relocation of affected residents. It is interesting to note that the Government of Trinidad and Tobago announced the approval of the project even in the face of such concerns and made it clear that environmental approval would be forthcoming. The government made the announcement of the approval of Train IV on June 5th, 2003 after a meeting of Cabinet. Interestingly, the Chairman of the local environmental agency, with responsibility for approving the project, responded to the announcement by the government that it would not be pressured into granting environmental clearance for Train IV. The comments of the Chairman were contained in a newspaper article of June 7th, 2003 (Quash, 2003). What was even more significant is that the local environmental agency had in fact already granted environmental clearance on June 6th, 2003, a mere day after the announcement of the Cabinet decision to approve Train IV. This environmental clearance has been the subject of great controversy as the locals who are claiming to be affected by the activities of Atlantic LNG are alleging that none of their concerns was addressed in the terms and conditions of the environmental clearance.

Environmental Protection and Economic Development: Prospects of Reconciliation

Having seen the great chasm that exists between the developed and developing world, it is not surprising that the underlying force for poverty alleviation and modernization was seen as economic development, loosely defined as the process of growth and change aimed at raising people's living standards (Smith, 1994). It soon became apparent that growth in a nation's wealth did not translate into substantial benefits for the poor. This realization led to the emergence of a new definition of development which emphasized poverty alleviation and inequality reduction (Jazairy, 1992). Economic growth was not supplanted but rather placed within a framework of social objectives (George, 1988). At the time of the emergence of this definition of development, a third was also being postulated. This definition included the satisfaction of the basic needs dimension of the second definition but went on to include elements of emotional, spiritual and political needs. Todaro, in articulating one of the most concise summaries of this wider perception of development, sees it as encompassing three core values: first, the satisfaction of basic physiological needs; second, self-esteem; and third, freedom from the social servitudes of men to nature, ignorance, other men, misery, institutions and dogmatic beliefs (1992). Today the notion of human development is increasingly being seen as the ultimate goal of economic development, with the objective of development not being to produce more goods but rather to improve the capabilities of people to lead full, productive and satisfying lives (Griffin and Mckinley, 1994). The human development approach has numerous advantages. It contributes directly to the well-being of people; it builds from a foundation of equality of opportunity; it creates a more equitable distribution of the benefits of development; it enables the linkages between the various types of investment in people to be fully exploited; and it takes advantage of the complementarities between human and physical capital (Griffin and Mckinley, 1994). Development is seen to be about people and not roads, machines and infrastructure. Poverty alleviation demands full participation by people in the developmental process. This participation must grow out of respect for human life, dignity and democratic values (Jazairy, 1992). In spite of the movement towards developmental objectives that would reflect the state of people and the concern with poverty in the developing world, there have been certain trends commencing in the 1980s that have forced a change in outlook.

Reviews of the different perspectives on development (Persadie, 2002) illustrate that they completely ignore environmental considerations in their developmental objectives, largely because countries "had not yet recognized their capacity to degrade [the environment] irreversibly on a large scale" (Weiss, 1992). This is not to say that widespread environmental destruction has not continued after having learnt of the deleterious effects of this type of behavior. Development in most countries is based on natural resource exploitation and, according to the major theoretical offerings on development, environmental degradation, as manifested by deforestation, pollution, and unmeasured exploitation, equals economic progress *á la* West. This type of development occurs despite the recognition that "population growth and rising production and consumption cannot be sustained forever in a finite world" (Ekins, 1996).

Countries in the initial stages of economic development cause severe environmental degradation "up to the point at which they acquire sufficient wealth to generate both a demand to protect the environment and the resources with which to do it. At this point, free trade practices promote environmental protection" (Weiss, 1992). This, however, reiterates the modernist approach to development and ultimately justifies development at the inevitable risk of environmental degradation. A shift away from the modernist impersonal aspect of development toward one based on "direct human connections can help create the notion of sustainable development as a fundamental human value" (Banuri, 1990).

Green development thinking offers an alternative, non-deterministic view to traditional development models. As mentioned above, the current development process is the cause of many of the problems faced by society today. The green view proposes that "the concept of development must be defined in human terms if it is to have any relevance at all" (Freiburg and Hettne, 1985). This developmental strategy views the relationship between man and nature as integrally interdependent, and does not view the environment simply as resources to be exploited for capital gains. They have replaced the concept of human mastery over nature with the idea of partnership, an idea alien to current development paradigms. The goal, however, "...is not to preserve nature while forgetting men, but rather to manage nature in such a way that the men of our generation and all future generations will be ensured of the possibility of developing" (Freiburg and Hettne, 1985). This, of course, mirrors the basic tenet of sustainable development, which does not espouse the need to do away with the 'traditional' aspects of society, but rather seeks to incorporate them into a new development model.

Given its fairly radical perspective, the Green view has no models to emulate. According to Freiburg and Hettne,

> A 'backward' country should not look for the image of its own future in the 'advanced' country, but in its own ecology and culture. ***There is no universal path to development. Each society must find its own strategy*** (1985, p. 220) (Emphasis added).

The success of any green development model is, therefore, dependent on the integration of the cultural identity of the country involved and the (cultural) appropriateness of technologies introduced. It calls for alternative technology, which is "small scale and compatible with the local culture and the local eco-system..." (Freiburg and Hettne, 1985).

Like any other theory, there must be a theoretical construction based on structural and historical conditions for the formulation of the Green perspective; the world-system approach offers a good framework from which to work, but it must assume a more voluntaristic and less deterministic direction. Further, the unit of analysis is radically different to those of the dominant perspectives. From a Green view, human beings or small communities of human beings are the ultimate actors. As Freiburg and Hettne offer, "The tribes and nations of the world are much more basic units of development, because they allow for the forging of a genuine consensus between their members" (1985).

This course of development lends itself to criticism, as it would appear to be a "utopian form of anti-modernism—utopian in the double sense of being a visionary panacea and unrealizable" (Freiburg and Hettne, 1985). However, Freiburg and Hettne suggest that there is need for realistic utopianism—radical solutions founded on critical analysis of the strength of various anti-systemic forces (1985).

It has already been illustrated that economic growth has not eradicated poverty from the developing world nor improved the health of the environment (Davidson, 1992). Harrison has pointed out that while many developing countries achieved rapid economic growth in the 1960s, the poorest 30 to 60% of the population experienced unchanging living conditions (1980). Therefore, growth at any price cannot provide a solution to the human and environmental problems of the developing world.

Even the World Bank has belatedly embraced the concept of sustainable development. Barber Conable, President of the World Bank, has admitted "that developing nations and development institutions have learned that sound ecology is not only good economics; it is essential for survival. It can be compatible with growth" (1991). Thus, the World Bank has claimed that up to 1991, more than 1,500 environmental components were added to power, transportation, industry and agriculture projects, with a few projects being implemented to improve soil conservation, to manage forests and rangelands, to prevent desertification, protect biological diversity and conserve water resources and fisheries (Conable, 1991).

Today there are numerous warnings of the need to ensure sound environmental protection in the developing world, yet the issue is afforded low priority. The low priority can be arguably ascribed to the presence of the dominant concern in developing nations of using economic development as a means to eradicating the social inequities that characterize these countries. The ultimate goal always seems to return to the alleviation of poverty. It is becoming clearer that the goals of economic development as the mechanism for poverty alleviation are in conflict with the aspirations of sound environmental management.

The major revolution in economic thinking emerged in the 1980s with the postulation of the economic development theory of sustainable development. Sustainable development received its most famous endorsement by the World Commission on Environment and Development (WCED) under the chairmanship of Gro Harlem Brundtland. Sustainable development recognizes that development must be geared towards meeting the basic needs of the vast numbers of poor in the developing world. It sees this happening by a process of change in which "the exploitation of resources, the direction of investments, the orientation of technological development, and institutional change are all in harmony and enhance both current and future potential to meet human needs and aspirations" (WCED, 1987). In June 1992, the Rio Earth Summit declared that "the right to development must be fulfilled so as to equitably meet developmental and environmental needs of present and future generations." The notion of sustainable development was significantly modified in Johannesburg in 2002. To paraphrase:

> We commit ourselves to undertaking concrete actions and measures at all levels and to enhancing international cooperation, taking into account the Rio principles, including *inter alia*, the principle of common but differentiated responsibilities as set out in principle 7 of the Rio Declaration on Environment and Development. These efforts will also promote the integration of the three components of sustainable development, economic development, social development and environmental protection as interdependent and mutually reinforcing pillars. Poverty eradication, changing unsustainable patterns of production and consumption and protecting and managing the natural resource base of economic and social development are overarching objectives of, and essential requirements for, sustainable development... The gap between developed and developing countries points to the continued need for a dynamic and enabling international economic environment supportive of international cooperation, particularly in the areas of finance, technology transfer, debt and trade and full and effective participation of developing countries global decision-making, if the momentum for global progress towards sustainable development is to be maintained and increased... (UN, 2002, pp. 8-9).

Despite the heady words contained in the evolving definition of sustainable development, the case analysis of Trinidad and Tobago paints a picture of "business as usual" with the environment continuing to be treated as political "window dressing". The environment receives public accolades in speeches both nationally and internationally and private abuse during the exercise of the executive decision making apparatus.

The major manifestation of the conflict between development and the environment can be found in the allocation of financial resources. Having regard to the reality that the state is, by and large, the driver of sound environmental policies, it stands to reason that the state would be required to bear the brunt of the financial investment in a sound and sustainable environment. However, the reality is that the state has limited resources and various priorities, and it is hardly surprising that the environment is of low or limited priority. The other significant dimension to the conflict between the environment and development can be found in the attitude towards natural resources. A key part of the environment is natural resources. What emerges without much contradiction is the trend that suggests that not only is the attitude to the environment manifested in the allocation of financial resources but it is also present in the approach to natural resources which are viewed in one of two ways: either as commodities to be exploited in a manner designed to maximize income or as hindrances to income earning activities. The sad reality is that the perception in developing countries is often that the environment is just capital waiting to be banked or obstacles to be summarily dismissed. Current models of development in developing economies will not allow for the development of sound international environmental management. Developing countries now have to adopt a different economic development paradigm that will seek to include the environment. It should assume a more holistic approach, seeking to achieve the Chinese principle of *feng shui*. In addition to self-sufficiency, there would also be an inherent appreciation for the value of nature, culture and the environment. There would also be the political empowerment of marginalized groups in society, such as women and indigenous groups. This would lead to a fully integrated society where there would be sustainability in the social, cultural, political economic and environmental realms.

Endnotes

1. Classification of economies. For operational and analytical purposes, the World Bank's main criterion for classifying economies is gross national income (GNI) per capita. Every economy is classified as low income, middle income (subdivided into lower middle and upper middle), or high income. The use of the term is convenient; it is not intended to imply that all economies in the group are experiencing similar development or that other economies have reached a preferred or final stage of development. Classification by

income does not necessarily reflect development status. A short history—in general discussions in Bank reports, the term "developing economies" has been used to denote the set of low and middle-income economies. Bank publications with notes on the classification of economies state that the term "developing economies...does not imply either that all the economies belonging to the group are actually in the process of developing, nor that those not in the group have necessarily reached some preferred or final stage of development." Economies are divided according to 1999 GNI per capita, calculated using the World Bank Atlas method. The groups are: low income, $ 755 or less; lower middle income, $ 756-$2,995; upper middle income, $ 2,996-$ 9,265; and high income, $ 9,266 or more. For purposes of this paper, references to developing countries are deemed to be references to low and middle income countries and references to developed countries shall be construed as references to high-income countries (World Bank, 2001). Figures are quoted in US dollars.

2. Figures were available for only four developed countries.
3. Chronically undernourished people are defined as those people whose estimated daily energy intake over a year falls below that required maintaining body weight and support light activity.
4. The survey years for countries for which data was obtained were not uniform.
5. Unemployment figures were derived from the World Bank, 2001. The ILO defines the unemployed as persons above a certain age (generally taken to be 15 years) available for work, not in paid employment or self-employed and seeking work (ILO, 1995). "Unemployment is defined as follows in the Resolution concerning statistics of the economically active population, employment, unemployment and underemployment, adopted by the Thirteenth International Conference of Labour Statisticians (Geneva, 1982):
    1. The 'unemployed' comprise all persons above a specified age who during the reference period were:
        (a) 'without work', i.e. were not in paid employment or self-employment, as defined in paragraph 9;
        (b) 'currently available for work', i.e. were available for paid employment or self-employment during the reference period; and
        (c) 'seeking work', i.e. had taken specific steps in a specified reference period to seek paid employment or self-employment" (ILO).
6. It should be noted that figures for radio ownership were few.

7. The figures for percentage of paved roads for developing and least developed countries were limited.
8. The foreign exchange rate used is US$ 1.00=TT$ 6.30.
9. For such a thorough examination one would have to collate and calculate information from the PSIP as well as data on current expenditure, not simply for monies allocated to the environment *per se*, but for agencies with environmental responsibilities
10. Unless otherwise stated, the information from this section is derived from a personal interview with Mr. Narine Lackhan, the Director of Forestry, the Forestry Division, the Ministry of Agriculture, Land and Marine Resources, conducted on 25 July 2000 at the Forestry Division.
11. The irony of this, though, is that these honorary Game Wardens are themselves hunters and may not necessarily enforce the law as is required.
12. While an EIA was undertaken, it was superficial and did not take into account several basic repercussions that would result from such a project, such as the downstream effects. Nor did it even include an oceanographic study to determine the feasibility of this project.
13. In the Matter of the Environmental Management Act, 2000 and the Certificate of Environmental Clearance Rules, 2001 Between Talisman (Trinidad) Petroleum Limited [Appellant] and the Environmental Management Authority [Respondent] No. EA3 of 2002.

References

Banuri, T. (1990). "Development and the Politics of Knowledge: A Critical Interpretation of the Social Role of Modernization", in F. Marglin and S. Marglin (eds.), *Dominating Knowledge: Development, Culture, and Resistance*. Oxford: Clarendon Press, pp. 29-72.

Bhaskara, H. (1989). *Against All Odds: Breaking the Poverty Trap*. London: Panos.

Bello, W., Cunningham, S. and Rau, B. (eds.) (1994). *Dark Victory: The United States, Structural Adjustment and Global Poverty*. London: Pluto Press.

Bursztein, S. (1989). *Energy Metabolism, Indirect Calorimetry and Nutrition*. Maryland: Williams & Wilkins.

Carajas: Overview Enough Iron Ore for the Next Five Hundred Years. (1998). Retrieved 31 July 2003, from http://www.southamericavacationguide.com/Carajas2527_Overview.html.

Chackhan, G. (2000, May 13). "Don't Destroy our Natural Assets", *Saturday Express.*

Chege, N. (1995). "Homelessness Remains a Problem", in L. Brown, N. Lenssen and H. Kane (eds.), *Vital Signs: The Trends that are Shaping Our Future 1995-1996.* London: Earthscan, pp. 142-143.

Colclough, C. (1993). *Educating all the Children: Strategies for Primary Schooling in the South.* Oxford: Clarendon Press.

Conable, B. (1991). "Ushering in the Era of Sustainable Development", in M.K. Tolba and A.K. Biswas (eds.), *Earth and Us: Population-Resources-Environment and Development.* Oxford: Butterworth-Heinemann, pp. 32-36.

Dasgupta, P. (1993). *An Inquiry into Well-Being and Destitution.* Oxford: Clarendon Press.

Davidson, J., Myers, D. and Chakraborty, M. (1992). *No Time to Waste: Poverty and the Global Environment.* Oxford: Oxfam.

De, Coteau and Gangar, A. (2000, May 16). "Energy Sector to Blame for Pollution", *Trinidad Express Politics.* Retrieved 17 May 2001, from http://209.94.197.2/may00/may16/politics.htm.

Economic and Social Commission for Asia and the Pacific (1992). *The Asian and Pacific Input to the United Nations Conference on Environment and Development Brazil, 1992.* New York: United Nations.

Ekins, P. (1996). "Limits to Growth and Sustainable Development: Grappling with Ecological Realities", in K. Jameson and C. Wilber (eds.), *The Political Economy of Development and Underdevelopment.* New York: McGraw-Hill Inc., pp. 53-69.

Environmental Management Authority (EMA). (1998). *National Environmental Policy.* Port of Spain, Trinidad: Government Printery.

Environmental Problems of Afghanistan (1972). Report submitted to the United Nations Conference on the Human Environment.

Freiberg, M. and Hettne, B. (1985). "The Greening of the World: Towards a Non-deterministic Model of Global Processes", in *Development as Social Transformation: Reflections on the Global Problematique.* London: Hodder and Stoughton, pp. 204-270.

George, V. (1988). *Wealth, Poverty and Starvation: A World Perspective.* Hamel Hempstead: Wheatsheaf Books.

Griffin, K. and Mckinley, T. (1994). *Implementing a Human Development Strategy.* Basingstoke: MacMillan.

Grindle, J. (1992). *Bread and Freedom: Basic Human Needs and Human Rights.* Dublin: Trocaire and Gill-McMillan.

Gupta, A. (1988). *Ecology and Development in the Third World*. London: Routledge.

Harrison, P. (1980). *The Third World Tomorrow*. Harmondsworth: Penguin.

Hassan, S. (1991). *Environmental Issues and Security in South Asia*. Oxford: Nuffield Press.

International Labour Organisation (ILO). (1995). *1995 Yearbook of Labour Statistics*. Geneva, ILO.

International Labour Organization. *Unemployment*. Retrieved 17 July 2002, from http://laborsta.ilo.org/applv8/data/c3e.html.

International Rivers Network (IRN). (2002, October). *Tehri Dam*. Retrieved 31 July 2003, from http://www.irn.org/programs/india/021022.tehrifactsheet.pdf

Jazairy, I., Alamgir, M. and Panuccio, T. (1992). *The State of World Rural Poverty*. London: Intermediate Technology Publications.

Kent, G. (1984). *The Political Economy of Hunger: The Silent Holocaust*. New York: Praeger.

King, F.S. and Burgess, A. (1993). *Nutrition for Developing Countries*. Oxford: Oxford University Press.

Meredith, M. (2000a, May 29). "Big Buck$ Invade the Bay." *Daily Express*.

Meredith, M. (2000b, May 29). "Government Breaks Own Rules: Mangrove Makes Way for $m Waterfront", *Daily Express*.

Meredith, M. (2002, November 10). "Talisman's Tour: An Explosive Issue", *Sunday Express*, p. 15.

Ministry of Agriculture, Land and Marine Resources, Trinidad and Tobago. (1999a). *1998 Annual Administrative Report of the Forestry Division*. Port of Spain, Trinidad: Government Printery.

Ministry of Agriculture, Land and Marine Resources, Trinidad and Tobago. (1999b). *1997 Annual Report of the Forestry Division*. Port of Spain, Trinidad: Government Printery.

Ministry of Energy and Energy Industries (MOEEI), Trinidad and Tobago. (1992). *A Draft Energy Policy for Trinidad and Tobago: A Green Paper* (Vols. 1 and 2). Port of Spain, Trinidad: Government Printery.

Ministry of Energy and Energy Industries (MOEEI), Trinidad and Tobago. (1994). *Draft Policy on Energy* (Vols. 1 and 2). Port of Spain, Trinidad: Ministry of Energy.

Ministry of Energy and Energy Industries (MOEEI), Trinidad and Tobago. (1998). *Green Paper for Proposed Energy Policy for the Republic of Trinidad and Tobago*. Port of Spain, Trinidad: Government Printery.

Ministry of Energy and Energy Industries (MOEEI), Trinidad and Tobago. (2000, May 28). "Overview: T&T oil Production", *Sunday Express.*

Ministry of Energy and Natural Resources, Trinidad and Tobago (1986). *Draft. An Energy Policy for Trinidad and Tobago.* Port of Spain, Trinidad: Ministry of Energy.

Ministry of Finance, Trinidad and Tobago (1997). *Medium Term Policy Framework 1998-2000.* Port of Spain, Trinidad: Government Printery.

Ministry of Planning and Development, Trinidad and Tobago (1999). *Public Sector Investment Programme 1999/2000.* Port of Spain, Trinidad: Government Printery.

Ministry of Planning and Development, Trinidad and Tobago (1997). *Public Sector Investment Programme 1998.* Port of Spain, Trinidad: Government Printery.

Ministry of Planning and Development, Trinidad and Tobago (1996). *Public Sector Investment Programme 1997.* Port of Spain, Trinidad: Government Printery.

Mugabe, R. (1991). "Environmental Concerns in the Third World", in M.K. Tolba and A.K. Biswas (eds.), *Earth and Us: Population-Resources-Environment and Development.* Oxford: Butterworth-Heinemann, pp. 10-13.

Pargal, S. and Mani, M. (2000). "Citizen Activism, Environmental Regulation, and the Location of Industrial Plants: Evidence from India", *Economic Development and Cultural Change,* 18(4): 829-847.

Patel, A.S. (1991). "Evaluation of Development Plans and Strategies for the Reduction of Poverty: The Case of India", in P. Henry (ed.), *Poverty, Progress and Development.* London: Kegan Paul International, pp. 203-227.

Pearce, D.W. and Warford, J.J. (1993). *World without End: Economics, Environment, and Sustainable Development.* Oxford: Oxford University Press.

Persadie, N. (2002). "Environmental Protection and Economic Development: Prospects for the Building of an International Environmental Legal Regime", Unpublished Master's Thesis, University of the West Indies, St. Augustine, Trinidad.

Pieterse, J. (2000). "After Post-development", *Third World Quarterly,* 21(2): 175-191.

Port Authority of Trinidad and Tobago (PATT), Portside News Weekly. (2000a, May 3). "Port of Spain Continues to Grow...from the Sea", *Daily Express.*

Port Authority of Trinidad and Tobago (PATT), Portside News Weekly. (2000b, May 21). "Status Report: Dredging of the Port of Spain Harbour", (21 May 2000b). *Sunday Express.*

Quash, C. (2003, June 7). "Atlantic LNG Train IV must Meet EMA standards", *Saturday Express*, p. 4.

Smith, C. (1994). *Economic Development, Growth and Welfare.* Basingstoke: MacMillan.

Some Aspects of Environmental Degradation and its Control in India (1972). Report Submitted to the United Nations Conference on the Human Environment.

The South Commission (1990). *The Challenge to the South: The Report of the South Commission.* New York: Oxford University Press.

Todaro, M.P. (1992). *Economics for a Developing World.* London: Longman.

Tourism Industrial Development Company (TIDCO) (1998). *Strategic Plan 1998-2000.* Port of Spain, Trinidad: TIDCO.

Tourism Industrial Development Company (TIDCO) (1999). *Trinidad and Tobago Export Directory.* Port of Spain, Trinidad: Government.

Tolba, M. and El-Kholy, O.A. (eds.). (1993). *United Nations Environment Programme, The World Environment 1972-1992: Two Decades of Challenge.* London: Chapman & Hall.

United Nations (2002). *Report of the World Summit on Sustainable Development.* A/CONF.199/20.

United Nations Conference on Environment and Development (UNCED) (1992). *Overview of National Reports.* Items 23 & 27. A/CONF.151/PC.98.

United Nations Development Programme (1994). *Human Development Report 1994.* New York: Oxford University Press.

United Nations Development Programme (1996). *Human Development Report 1996.* Oxford: Oxford University Press.

United Nations Development Programme (2001). *Human Development Report 2001: Making New Technologies Work for Human Development.* New York: Oxford University Press.

United Nations Centre for Human Settlements. *100 Million Homeless in World: Most are Women and Dependent Children.* Retrieved 17 July 2002 from http://www.un.org/Conferences/habitat/unchs/press/women.htm.

United Nations Centre for Human Settlements and The World Bank (1993). *The Housing Indicators Programme, Volume III: Preliminary Findings.* Nairobi: United Nations Centre for Human Settlements.

United Nations Centre on Human Settlements (1999a). *Statistical Annexes to the Global Report on Human Settlements 2001*, Table A-4. Retrieved 17 July 2002 from http://www.unchs.org/habrdd/Table%20A_4.doc.

United Nations Centre on Human Settlements (1999b). *Statistical Annexes to the Global Report on Human Settlements 2001*, Table B-3. Retrieved 17 July 2002 from http://www.unchs.org/habrdd/TABLE%20B_3%20.doc.

Weiss, E. (1992). "Environment and Trade as Partners in Sustainable Development: A Commentary", *The American Journal of International Law,* 86, 718-735.

World Bank (2001a). *World Development Indicators 2001*. (CD-ROM).

World Bank (1992). *World Development Report 1992: Development and the Environment*. New York: Oxford University Press.

World Bank (1995). *World Development Report 1995: Workers in an Integrating World.* New York: Oxford University Press.

World Bank (2000). *World Development Report 1999/2000: Entering the 21st Century.* New York: Oxford University Press.

World Bank (2001b). *World Development Report 2000/2001: Attacking Poverty.* New York: Oxford University Press.

World Commission on Environment and Development (WCED) (1987). *Our Common Future.* Oxford: Oxford University Press.

Yarbrough, B. and Yarbrough, R. (1994). *The World Economy: Trade and Finance.* Fort Worth, Texas: The Dryden Press.

Evaluation of Effective Microorganisms (EM) in Solid Waste Management

5

V. Sekeran, C. Balaji, T. Bhagavathi Pushpa

Organic manures are derived from plant, animal and human residues. In addition to supplying principal elements to the soil, they promote microbial activity in the soil and improve its structure, aeration, and water holding capacity, which in turn improves the soil capabilities to respond to inputs. This paper deals with the composting of kitchen wastes using effective microorganisms (EM), which results in a higher decomposition of organic matter and no odor during process. This produces a high quality compost, which contributes toward sustainable development.

Organic Matters, Compost, Effective Microorganisms (EM)

A major problem facing municipalities throughout the world is the treatment, disposal and/or recycling of solid wastes. Generally solid waste from a municipality consists of biodegradable organic materials. At the present time, there are a number of methods being used to dispose of the solid wastes in a landfill. Although there are many methods used, it requires the selection of the correct method focusing on efficient and environmentally safe disposal. New technology is being produced to assist the organic waste treatment, conforming to strict environmental regulations. One of those new technologies being proposed is the use of effective microorganisms (EM).

The technology of EM was developed during the 1970s at the University of Ryukus, Okinawa, Japan (Sangakkara, 2002). Studies have suggested that EM may have a number of applications including agriculture, livestock, gardening and landscaping, composting, bioremediation, cleaning septic tanks, algal control and household uses (Higa and Chinen, 1998) .

EM is a mixture of groups of organisms that has a reviving action on humans, animals and the natural environment (Higa, 1995) and has also been described as a multi-culture of coexisting anaerobic and aerobic beneficial microorganisms (EM Trading, 2000).

The main species involved in EM include:

- Lactic acid bacteria: *Lactobacillus plantarum, L. Casei, Streptococcus lactis.*
- Photosynthetic bacteria: *Rhodopseudomonas palustrus, Rhodobacter spaeroides.*
- Yeasts: *Saccharomyces cereuisiae, Candida utilis.*
- Acitinomycetes: *Streptomyces albur, S. griseus.*
- Fermenting fungi: *Aspergillus oryzae, Mucor hiemalis* (Diver, 2001).

The basis for using these EM species of microorganisms is that they contain various organic acids due to the presence of lactic acid bacteria, which is a strong sterilizing compound and suppresses harmful microorganisms and enhances decomposition of organic matter. Also they have the ability to suppress disease inducing microorganisms such as Fusarium, which occur in continuous cropping programs.

Materials and Methods

Sample

The sample that was used for composting was collected from the college campus, which includes the wastes from the canteen and hostels.

Activating the EM

EM is available in a dormant state and requires activation before application. Activation involves the addition of 20 liters of water and 2 kilograms of jaggery (pure cane sugar) to 1 liter of dormant EM. Pour the mixture into a clean airtight plastic container with no air left in the container. Store the container away from direct sunlight at ambient temperatures for 8 to 10 days. Release the gas everyday until fermentation complete.

During the period of activation, a white layer of *Acitinomycetes* forms on the top of the solution accompanied by a pleasant smell. The pH is also a determining factor; the pH of the EM should be below 4.0.

Experimentation

A solution was prepared by mixing 30 liters water and 1 liter activated EM in a plastic bucket and about 15 liters were sprayed on the clean 3 × 1 meter composting site. A 15 cm thick base layer containing animal waste was spread over the site and the activated EM was sprayed over this layer. A second layer of solid waste, about 30 cm thick, was spread over the previous and also sprayed with the EM solution. This layering process was repeated to a height of about 100 cm. The stack was completed with a final layer of 5 cm of animal dung. The entire unit was kept moist by spraying it with activated EM solution at regular intervals. After about 25 to 30 days the volume of bed had dropped substantially and a sweet-smelling white mold appeared on the biomass. At this point, the finished compost was collected and sieved. The compost were analyzed for pH, organic content, nitrogen (N), phosphorous (P), and potassium (K).

Results and Discussion

The parameters (nutrients carbon, nitrogen, phosphorous and potassium, and pH) of the fresh waste and the compost are presented in Tables 1 and 2.

Table 1: Parameters of the Fresh Waste

S. No.	Parameters	% by Dry Mass
1.	Carbon (C)	32.16
2.	Nitrogen (N)	0.98
3.	Phosphorous (P)	1.02
4.	Potassium (K)	0.402
5.	C: N	28: 1
6.	pH	7.4

Table 2: Parameters of the Composted Waste

S. No.	Parameters	% by Dry Mass
1.	Carbon (C)	30.05
2.	Nitrogen (N)	1.21
3.	Phosphorous (P)	0.63
4.	Potassium (K)	0.40
5.	C: N	26: 1
6.	pH	8.4

The carbon content of the wastes decreased during composting indicating a higher mineralization of organic matter. However, the nitrogen content increased in the process of composting. This shows that the increased microbial activity continues in the casts and results in an increased mineralization rate of organic nitrogen and consequent further increase in concentration of NH^{4+}.

In the present study the carbon and nitrogen (C : N) ratio measurement provided an indication of the degree of decomposition. The C : N ratio of the wastes was higher before composting than after. Enhanced organic matter decomposition in the presence of effective microorganisms results in lowering the C : N ratio. In general, a low C : N ratio accelerates the rate of decomposition but may cause a loss of nitrogen as ammonia gas and a rapid depletion of the available oxygen supply, leading to foul smelling conditions. But EM controls the foul smell and the process is free of odors.

During composting, the organic carbon is lost as CO_2 and total nitrogen increases as a result of carbon loss. The final nitrogen content of compost is dependent on the initial nitrogen present in the waste and the extent of decomposition (Crawford, 1983).

There was a shift in pH from the initial condition toward an acidic condition. The occurrence of acidic conditions may be attributed to the bioconversion of the organic material into various intermediate types of organic acid and higher mineralization of the nitrogen and phosphorous into nitrites/ nitrates and orthophosphate respectively.

Conclusions

The study revealed that:

- Kitchen wastes provided a better environment for the effective microorganisms (EM) to grow and produced a higher quality of compost.
- Organic matter helps in soil management for sustainable cultivation of any crop.
- The adaptation of effective microorganisms (EM) leads to detoxification of our landfills, decontamination of our environment and promotes highly sustainable, closed-cycle agricultural and organic waste treatment.

References

Crawford, J.H. (1983). "Review of Composting", *Process Biochemistry,* 18: 14-15.

Diver, S. (2001). "Nature Farming and Effective Microorganisms", Retrieved from *Rhizosphere II: Publications, Resource Lists and Web Links from Steve Diver* http://ncatark.uark.edu/~steved/Nature-Farm-EM.html.

Higa, T. (1995). *What is EM Technology*. Okinawa, Japan: University of Ryukyus, College of Agriculture.

Higa, T. and Chinen, N. (1998). *EM Treatment of Odor, Wastewater, and Environmental Problems*. Okinawa, Japan: University of Ryukyus, College of Agriculture.

Higa, T. and Wood, M. (n.d.). *Effective Microorganisms for Sustainable Community Development: A National Case Study of Cooperative and Co-Prosperity in North Korea for the Preservation of Environmental, Agricultural, Economic, and Cultural Integrity.* Abstract retrieved from http://emtrading.com/em/htmlpapers/nkoreaab.html

Sangakkara, U.R. (2002). *The Technology of Effective Microorganisms: Case Studies of Application.* Cirencester, UK: Royal Agricultural College.

The Social and Ethical Aspects of Nuclear Waste

6

Alan Marshall

Nuclear waste management seems to exist in a perpetual state of crises. For 50 years the nuclear states of the world have fought, and generally lost, the battle to deal with the nuclear waste problem. Worldwide, there is a growing acknowledgment within industry and government that social and ethical issues are just as important as technical issues when developing safe programs for nuclear waste management. This paper is a review of some of the outstanding social and ethical issues that are influencing discussions on nuclear waste management around the world.

Social Equity in Nuclear Waste Management

There are many ways that nuclear waste management has the potential to be socially inequitable: burdening certain groups of society with more than their fair share of risks and costs. The following sections outline salient social themes that have emerged as the nuclear nations in the world attempt to deal with nuclear waste:

Nuclear Stigma

According to work by Slovic, Layman and Flynn (1993, p. 64) nuclear waste can be regarded as the top neighbor from hell, ranking higher than oil refineries, chemical plants, garbage dumps and even nuclear power stations as the most undesirable facility to live beside.

The aversion to things nuclear, including nuclear waste, is often referred to as nuclear stigma and it has a number of possible effects: economic, social, political, cultural and psychological. With regard to the last of these, while there may be a case to state that the people of nuclear host communities are active in the construction of a positive nuclear identity, it is apparent that some members of the public are concerned about the mental stress of living close to a nuclear site (or the prospect of the same) (Dunlap, Rosa, Baxter and Mitchell, 1993; Edelstein, 1988). In such circumstances, if nuclear waste managers are to take social issues seriously then may be they should consider the ideas brought out by the likes of Lois Wilson (2000, p. 87), and Wendy Oser and Molly Young Brown (1996) who suggest professional counseling in some form should be provided to local individuals or groups. Kristen Shrader-Frechette (1993) suggests also that giving citizens funding for education and health might alleviate this problem, as might delegating authority to monitor stress to the community itself. This would allow local people to have some degree of self-help capacity over their own psychological and stress problems.

Another type of stigma that may rear its head in the siting of radioactive waste facilities is that associated with moral stigma. Easterling and Kunreuther (1995, p. 137) indicate that the moral qualms that people feel toward nuclear weapons seem to have generalized to civilian nuclear power. And thence, to anything nuclear, such as the radioactive waste left over from nuclear weapons and nuclear power production. In this case, if a nuclear waste management facility goes against the morals of individuals, it is not only politically problematic, giving rise to resistance, but ethically problematic, asking people to live with a facility they find morally objectionable. As far as these people are concerned, it is flippant for nuclear waste facility planners to derail weapons/waste connections by indicating that they are only involved in the rear-end of the nuclear cycle, when so much of the waste was produced for military purposes.

Nuclear stigma has also been identified as having identifiably negative economic consequences. New industries may be reluctant to set up near nuclear waste facilities in fear that their products will suffer negative nuclear stereotyping (Great Britain, Parliament, House of Lords, Select Committee on Science and Technology, 1999, p. 43).

In the states of Nevada and Texas, for example, pre-emptive concerns were expressed regarding the reputations of the tourist and cattle industries when sites in these states were considered for nuclear waste facilities proposed by the U.S. Department of Energy (Brody and Fleishman, 1993, p. 117; Slovic and Flynn, 1991; Easterling and Kunreuther, 1993). Similarly agricultural communities in eastern Washington state were concerned that the establishment of a nuclear repository at Hanford would be seen as leading to the contamination of fruits and wines grown in the area, thereby causing a decline in the economy (Easterling and Kunreuther, 1995, p. 137).

It must also be noted that while some community members might welcome an influx of industrial activity into their local area, desperate to "attract any kind of economic growth" (Rosa, Dunlap and Kraft, 1993, p. 303), others may fear that such an influx may lead to "increased crime, increased cost of living", property devaluation and disruption of their livelihood (Nuclear Energy Agency, Radioactive Waste Management Committee, 2003).

There has been some indication that what stresses the public most about nuclear power and radioactive waste is the possibility of an accident (Rosa et al., 1993; Easterling and Kunreuther, 1995). This is confounded by the suspicion that the managers of radioactive waste will be secretive with regard to the public dissemination of information about accidents (Flynn, Slovic, Mertz and Toma, 1990).

Nuclear Oases

Radioactive waste is an intensely local issue. The waste has to be located somewhere, whether it is stored or disposed of, and some communities are going to live nearer to this spot than others.

Once upon a time, when the dangers of nuclear activities were not generally well-known, it was usually the case that certain nuclear host communities were very positive about their status. Living in a pre-nuclear stigma era, many host communities felt they were partaking in a beneficial and advanced technological industry that brought jobs and services to their area. Since the late 1960s, however, this unfettered optimism has been battered by the changing economics of nuclear power and faltering tolerance of anything nuclear.

In some nations, the United States and the United Kingdom amongst them, the nuclear industry is suffering a slow but observable decline. Nuclear host sites and their adjacent communities, however, might be labeled as *nuclear oases*; a term the U.K. social scientist Andrew Blowers uses to denote places of lively nuclear activity in a world gradually deserting the industry (Blowers, Lowry and Solomon, 1991). According to Blowers (1999), nuclear oases are peripheral communities, in so far as they tend to be remote, economically and politically marginal and environmentally degraded. Examples of such communities, suggests Blowers, might include Sellafield in England, Hanford in the United States, Dounreay in Scotland, and Cap de la Hague in France.

These localized nuclear communities, Blowers (1999) intimates, exist as sites of intense interest for the nuclear industry, the last strongholds of economic and technical survival against a changing world. Generally, though, and despite the nuclear interest, nuclear oases are sometimes sites of neglect as far as national economy and public profile is concerned. Burdened with remoteness, marginality and powerlessness and previous environmental degradation the above named communities exhibit "a relatively stable locational pattern as a declining industry is resisted in all but the nuclear oases" (Blowers, 1999, p. 242).

Blowers' idea of nuclear oases is supported by American social science work on nuclear waste. For example, the social scientists Douglas Easterling and Howard Kunreuther (1995) have observed that traditionally there are lower levels of resistence and protest to new nuclear facilities in regions of strong nuclear presence where the residents may be dependent upon the jobs that the nuclear industry brings. Through such common dependence, a community spirit of defensiveness against anti-nuclear protest becomes inscribed in the minds of much of the local people (Blowers, 1999).

Of course, not all people living in nuclear oases may be there because they work for the nuclear industry, and some within the industry may themselves be quite critical of it. This has prompted some to note that nuclear host communities exhibit certain schism with regard to nuclear resistance. As well as schisms within the community, it is quite probable that individuals and family units may exhibit schisms of resistance and non-resistance. When social scientist Brian Wynne (1996) was studying the communities around Sellafield in the United Kingdom, for example, he found that Cumbrian farmers not far from Sellafield:

> recognized their own indirect and sometimes direct social dependency upon the Plant—not only neighbors, but also close relatives of the hill farmers worked there. Thus, underlying and bounding their expressed mistrust of the authorities and experts, there was a counter-veiling deep sense of social solidarity and dependency—social identification with material kinship, friendship, and community networks which needed to believe that Sellafield was well controlled and its surrounding experts credible (p. 37).

Wynne doesn't believe such schisms represent an inability to decide upon the ultimate goodness or badness of the nuclear plant but as a considered strategy to tread between various allegiances and experiences (p. 43).

According to Buclet and Bouzidi (2003), who studied nuclear host communities in France, the presence of nuclear oasis communities strongly familiarized to nuclear power does not give rise to more gentle resistance to nuclear waste issues. However, it should be noted that this resistance in the French case is not necessarily community-based but involves the actions of activist groups from metropolitan centers away from the nuclear host communities. In their book *The International Politics of Nuclear Waste*, Blowers, Lowry and Solomon (1991) also acknowledged such a phenomenon in the United Kingdom and the United States, as does Sj ölander (2003) in the case of one of Sweden's proposed nuclear host communities. If such examples are to be trusted, then we should predict that all nuclear waste host communities are liable to garner increasing help and attention from formally organized protest groups and informally organized urban sympathizers.

Regional Justice

If nuclear facilities happen to be clustered in particular parts of a nation then radioactive waste can become a regional phenomenon, thereby giving rise to issues of regional environmental justice (or geographical equity, as some writers like to call it (Gowda and Easterling, 2000). This issue has been brought to light by a number of writers in a number of countries. Lois Wilson (2000), in regard to Canada, points out, for instance, regional injustices whereby the south produces nuclear waste while the north is focused as the future repository of it. She cannot offer any process to resolve this injustice but merely asks what is the best way to address "equitable distribution of costs, risks and benefits among regions" (p. 3).

Easterling and Kunreuther (1995, p. 35) also point out that an unequal relationship of regions is something keenly felt by western states in the United States. Eastern states, which have a greater population and a greater electricity use, have historically looked west when they are searching for sites for the long-term management of the waste. Anti-nuclear waste sentiment in the western states of Nevada and Utah has given rise to cries of regional injustice when it comes to the planned nuclear waste facilities at Yucca Mountain (Dunlap, Rosa, Baxter and Mitchell, 1993) and Skull Valley (Fahys, 2003). The governors of these states have repeated the complaint that they do not produce nuclear waste, so therefore they should not have to store it for those who do (Gerrard, 1996).

The Promise of Employment

If nuclear communities and nuclear regions are economically depressed and sometimes financially stricken then perhaps they should be quite pleased to host a radioactive waste facility since it may well offer up new employment opportunities. Atomic Energy of Canada Limited (AECL), for example, makes the point that when construction of the Canadian Shield disposal facility starts, "jobs will be created" (as quoted in Wilson, 2000, p. 40). The International Atomic Energy Agency (IAEA) is also on record as saying that employment opportunities associated with nuclear facilities can foster community acceptance of the facility (International Atomic Energy Agency, 2003).

The immediate response from those who receive such advice, as Wilson (2000) notes, is to question exactly who will get these jobs. Will it be local people pooled from the surrounding community, or will the jobs go to imported skilled workers from different regions of the country? If the latter is more probable than the former, then any social impact analysis must be critical of the claim that long-term waste management facilities decrease local unemployment.

What also has to be assessed is the variation of the grades and qualities of work available to the local workers compared to the imported workers. The local workers, if unskilled in the nuclear industry, are more liable to be given the low-paid jobs and, if Shrader-Frechette's (2001) research is to be trusted, they also may be far more likely involved in non-unionized, low-profile, dangerous work for which they are under-prepared and underpaid with respect to the risks. For these reasons, it has sometimes been expressed that the promise of jobs is not sufficient to garner community acceptance of nuclear waste (Great Britain, Parliament, House of Lords, Select Committee on Science and Technology, 1999, p. 43).

Coercion and Consent

In the case of nuclear waste planning, it is an accepted belief within social science circles that a facility that imposes risks on a community should be built only if the members of that community give their consent (Gowda and Easterling, 2000). But an important issue that emerges involves the way that a potential nuclear host community may be pressured into offering up their consent.

Many prospective facilities have come across stiff opposition when proposed by governmental or private bodies. Despite this, though, the resources and funds that nuclear resistance groups are able to muster compared to the nuclear industry and government is very small. Governments and business can inject funds into their side of the proposal to produce advertisements, campaigns, education projects, and so forth, all aimed at fostering a public opinion conducive to their plans. If consent is given within such an atmosphere of often subtle but perfectly legal coercion, then what is the ethical status of the facility?

Normally we would regard all players in technology and environment debates as rational and well-informed actors capable of making up their own minds. For instance, if a radioactive waste facility was planned in a disused metro station in central New York or London and then opposed by the local people, we'd regard the people as being quite rational and informed. But as Blowers and Shrader-Frechette have illustrated, the communities subjected to waste facility plans (and the workers who are promised jobs in these facilities) may be regarded as peripheralized communities and economically-disadvantaged workers, unable to access all the information they need, unable to access independent points of view, and unable to fully judge the economic benefits versus the radiological risk.

All this gives rise to what Shrader-Frechette (1991) and Wigley (Wigley and Shrader-Fechette, 1994) would call the *consent dilemma*: wherein the siting of nuclear waste facilities and the employing of nuclear waste workers requires the consent of those who are put at risk; yet those most able to give free, informed consent are usually unwilling to do so, and those least able to validly consent are often willing to do so because they are unaware of the dangers.

These problems then beg us to ask the following questions with regards to siting nuclear waste facilities.

- What is an adequate level of information and understanding for people to make a decision?

- Do all stakeholders have equal access to adequate information and assistance in understanding?
- Who should be in-charge of ensuring adequate and equally-accessed information and understanding?

Compensation

One way of dealing with many of the issues noted above is to enact some form of retributive justice, typically compensation, for the people and communities affected. This path in itself is fraught with problems. Kleindorfer et al. (1988), for example, have produced evidence that some people do not believe any amount of compensation makes up for living next to a radioactive waste site.

Shrader-Frechette (1993, p. 204) warns that the use of compensation confuses and upsets any notion of pure consent. Although many people would acknowledge we live in a complex political world where consent always has to be negotiated, the problem is that the disparities in negotiating strength might arise purely through well-financed interests employing misinformation and propaganda, something that has to be countered if the act of compensation is to be processed in an open and fair way.

As an example of nuclear waste financial compensation in action, consider the town of Eurojoki in Finland whose council accepted over 6 million Euros from the waste production and management company Posiva to site a repository near their community. The Eurojoki council did not put the question to referendum amongst its community members but used a Posiva-conducted poll (Posiva Oy, 1999)—that indicated 59% of the community might accept the repository—to make a decision on behalf of the community. This high level of acceptance, 59%, in Eurojoki may have been because the town was already host to two nuclear reactors and so the community members involved in these nuclear operations could have been quite accustomed to the risk of happily clawing economic benefits from the handling of nuclear materials. However, according to Jorma Jantunen, a critic of the Eurojoki nuclear project, Posiva was bombarding the community with an advertising campaign, served not to inform the community members about the project but to get them to be positive about it (Nuclear Energy Agency, Radioactive Waste Management Committee, 2002).

Alternatively, if Blowers (1999) is right, the community members of Eurojoki may be so economically dependent on the nuclear industry that they feel unable to resist further nuclear operations in fear of industry's declining future. Added to this, if we put store in the writings of Blowers (1999), Shrader-Frechette (1991), Dunion (2003), and others, it may be that community members not actively involved in current nuclear operations may have been socialized to accept the industry's view of the risks and benefits without having the intellectual and financial resources to assess and challenge these received views. When financial compensation is introduced in a form such as that offered by Posiva, it is likely that some will perceive the process as being somewhat morally corrupt (Oughton, Bay, Forsberg, Hunt, Kaiser and Littlewood, 2003, p. 35).

Gender and Risk Sensitivity

A general feminist critique would posit that a lot of environmental and technology policy is biased towards male interests and perpetuates a patriarchal society (Buckingham-Hatfield, 2000; Everts, 1998). As a possible example of the gendered nature of radioactive waste, the report to the 3rd COWAM Seminar (*History and Some Facts to Wellenberg*, 2002) indicates that only 41% of women polled in a potential repository site accepted the idea of a nuclear waste repository in their area compared to 52% of males. Other commentators, like Gregory and Satterfield (2002), have noted that woman have a greater degree of sensitivity to risk in various hazardous environmental projects. Undoubtedly, there are a myriad of reasons for such situations: the sensitivity of women as a social group to environmental issues due to their self-perceived social roles, the sensitivity of men as a social group to technical issues due to their jobs, the higher expectations within men that economic benefits will actually help them and their families compared to a lower expectation among women for the same thing.

Indigenous Issues

Many countries with historical settler-populations have laws maintaining the land rights and personal rights of indigenous communities. Some of these countries, for instance, the United States, Canada and Australia, have nuclear waste. In these countries it often happens that nuclear waste facilities are proposed in remote areas occupied by a high proportion of indigenous people or near to indigenous reservations. An added concern is that these communities are often peripherilized and economically disadvantaged (Fowler, Hamby, Rusco and Rusco, 1990). This is a recipe for deep social injustice based not only on regionalism and economic inequality but on ethnic issues as well. For instance, Lois Wilson (2000) in Canada noted that one representative of the Canadian indigenous community in a preliminary hearing said that he:

> represents fifty First Nation communities, inhabiting two-thirds of the Ontario land mass. Thirty-five of these communities do not have road access, twenty-five are not connected to the electric power grid, and none use nuclear power (p. 16).

In Canada, the responsible authorities have now at least recognized the necessity to incorporate indigenous concerns into radioactive waste management (Nuclear Energy Agency, Radioactive Waste Management Committee, 2003). Amongst the cited concerns of indigenous groups within targeted sites are the issues of maintaining access to water and land resources, protecting the quality of these resources, health and safety against accidents and pollution, protecting important historical and cultural sites, and sustaining and enhancing cultural and economic opportunities for community members.

NIMBYism

Negative public reactions to radioactive waste facilities are often construed as an operation of the NIMBY (Not-In-My-Back-Yard) syndrome. NIMBYism, under this interpretation, is the emotive, reactionary impulse of local citizens to a project they would probably agree with were it placed somewhere else. Some, like Rosa, Dunlap and Kraft (1993), feel that such

NIMBYism may just be the predictable result of the alienation that people feel to national decision-making processes, a natural response to their resignation that their views will not ever be considered.

According to some research, the whole concept of NIMBYism has little explanatory power when used to interpret the politics of managing and siting radioactive waste facilities. The NIMBY concept predicts that those people physically closest to any planned facility should be those most objecting to it, but when Krannich, Little and Cramer (1993) studied the phenomenon as applied to the Yucca Mountain repository in Nevada they found that opposition and concern are strongest in the communities farthest from Yucca Mountain.

Another theme that the faltering NIMBY concept predicts is that the arguments of opponents will be emotionally driven by fear and dread and that they will be lacking in technical sophistication. But according to Kraft and Clary (1993), who were studying repository-siting meetings, only 14% of those members of the public testifying made declarations of this kind. Emotive themes were present for only a relatively small number of those making statements; the vast majority did not appeal to emotionalism.

Kraft and Clary also repeat the idea forged by numerous previous studies that a great amount of public testimony from non-expert individuals and groups is of comparable technical sophistication to that of the experts (Martin, 1996).

After reviewing the way public acceptance of a facility is either forthcoming or not within various affected communities across the United States, Rosa et al. (1993) come to the conclusion that resistance to nuclear waste is so widespread that it does not conform to NIMBYism at all but to NIABYism: Not In Anyone's Backyard (p. 318).

Although NIMBYism is denounced by many project planners as the irrational knee-jerk reaction of technically unsophisticated locals acting out of self-interest, if we trust the research outlined above, it seems as though the quick and indiscriminate labeling of resistance as NIMBYism is but the knee-jerk reaction of politically unsophisticated project planners who themselves are reacting under self-interest. A number of works, like for instance that of Rabe (1994), Dunion (2003) and McAvoy (1999), would confirm this view.

Lack of Public Understanding

One of the concerns that arises from the side of the nuclear industry regarding nuclear waste management is that the public does not fully understand the technical issues at hand. This makes it impossible for the nuclear industry to garner full public acceptance of their plans.

This perceived public deficit of knowledge gives rise to what Alan Irwin and Brian Wynne label the *public ignorance* model of citizen participation. If only the public can be rescued from their ignorance, this model suggests, they would be freed of their irrational dread associated with nuclear operations. The public ignorance model, which advocates a form of public participation based upon education, has its roots in the presumption held by many scientists and technologists that the reason people do not fully trust the scientifically-proven point of view is because the public don't fully understand it. For example, Sundqvist (2002) says:

> There is a widely held image, in the rhetoric of decision makers, of lay people as uninformed, ignorant and fearful of the unknown. This image suggests that if the level of information is raised, lay people will accept the proposals from decision makers (p. 14).

Rosa et al. (1993) echo this point with regard to the 50 years of nuclear facility siting in the United States:

> The nuclear sub-government, then as now, was guided by the unshakeable belief that increased public understanding—the knowledge fix—would translate into support for nuclear technologies. All that was required was thoughtful public relations to convert the dull, scientific knowledge into interesting, convincing public knowledge (p. 77).

Susana Hornig Priest (Hornig Priest, Bonfadelli and Rusanen, 2003), drawing from her social studies of biotechnology, points out that any determined effort to use public relations to educate the public about controversial science and technology is prone to backfiring. Rosa et al. (1993, p. 315) have found that the same thing happens when the nuclear industry starts up campaigns aimed at using the media to disseminate information.

Transportation Issues

Within and outside of the industry, the transport of nuclear waste has been perceived as inherently riskier than its storage or disposal. The risk of such accidents has driven some writers to declare that waste transport should be regarded as the last resort (Nuclear Guardianship Project, 2002).

According to studies by Slovic et al. (1993), somewhere between 70% and 80% of people questioned in Nevada and California were convinced that railway and highway accidents were going to occur on route to any operating nuclear waste facility. The public perception of transportation as being a problem arises in part from the acknowledged dangers emerging from industry watchdogs, the media, and the industry itself. For instance, the Association of Electronic Journalists declares that "from 1971 to 1998, there were 1,936 accidents and incidents involving radioactive materials transport" (*Nuclear Shipping Accidents: Rare but Regular*, 2002).

When forecasting the transport problems of the proposed Yucca Mountain repository in Nevada, the U.S. Department of Energy (DOE) predicted there will be 100 accidents over the lifetime of the project (the State of Nevada predicts 400 accidents during the same period) (Wile and Cox, 2002). Most of these accidents would result in no, or negligible, harm to human health and the environment. However, Wile and Cox used published DOE figures to study what that agency calls a "moderate" accident. Wile and Cox concluded that under such an event:

- A small number of first responders may be fatally affected.
- Around 200 to 1,200 latent fatal cancers of nearby citizens would eventuate.
- Nearly 600 million dollars would be needed to clean up the contaminated area over a 14 month period.

In the event of a transport accident it is fairly certain that local fire, police and ambulance services might be among the first upon the scene. An ethical issue that must be investigated here is whether all the emergency personnel from the local communities that line the proposed routes of the transported radioactive waste should be trained in some way to deal with accidents that may involve that waste. If so, this will have ramifications concerning the security and financial regimes under which such training might be given.

Some people have argued that the transportation of waste is so dangerous that it should not be undertaken. The Nevada-based Citizen Alert group, for instance, points out that transportation massively increases all the risks associated with radioactive waste handling (*High Level Radioactive Waste Transportation Factsheet*, 2000). Physical, or *passive*, security, for instance, at stationary sites involves much more robust physical protection from human interference and natural disaster since the strength of buildings and earthworks that house stationary waste is greater than that achievable with mobile wastes. Nevada's Nuclear Waste Project Office confirm this when they declare that if transport casks were designed to protect the waste to the same degree as stationary facilities, they'd be too heavy to be transported (Nevada Nuclear Waste Project Office, 1999).

When it comes to *active* security, mobile radioactive waste cannot favorably compare to the stationary waste either, since the former does not have the police presence, and the emergency personnel, that regularly accompanies the latter. The Nuclear Information and Resource Service (NIRS) (Mariotte, 1998) also points out that mobile radioactive waste is more vulnerable to external factors than stationary waste since, as safe as we can get the transportation system, external factors (such as drunken drivers, weather extremes, traffic emergencies—all of which have caused accidents in radioactive transport in the past) cannot be eliminated.

Another important issue regarding transport of radioactive waste is whether the route should be openly declared. To discuss this particular issue necessitates an engagement with the never-ending balancing act of working with security concerns versus fairness/democratic concerns. To minimize the risk of terrorist action or theft, the usual approach is to keep the routes secret. To maximize the democratic impulse of people to know about threats to their health and their environment, the routes should be declared. This balance may be made more complex by acknowledging that some along the route are more concerned about nuclear stigma affecting property prices than about any health risk or environmental danger. Thus, under the rhetoric of fairness, there may be social pressure (and also political back-up) for the routes to remain unnamed (Gawande and Jenkins Smith, 2001).

Public Participation Issues

Projects to manage nuclear waste involve a series of important and perhaps irreversible decisions. In many nations it is generally thought that these decisions should reflect a certain amount of public involvement (Kraft, Rosa and Dunlap, 1993, p. 11).

Not all people are of the opinion that nuclear waste is an issue worthy of extensive public consultation. Many technocrats believe that when it comes to siting radioactive waste, it is unlikely that everybody can have their desires catered for, but that this situation shouldn't stop the government from making a decision in favor of the interests of the majority (Great Britain, Parliament, House of Lords, Select Committee on Science and Technology, 1999).

Public participation has come very slowly to nuclear issues. The secrecy of the nuclear industry and its strategic importance regarding security and military affairs has encouraged this. In some nuclear nations this modus operandi of secret operations continues unabated. There has also been the attitude that such complex technical issues should be left to those experts trained in nuclear science and technology management.

In those nations that claim to have strong democratic governments, however, public participation is gradually becoming more extensive and more intensive. The most rudimentary form of public participation is the breaking down of the secrecy barriers just mentioned. This form of public participation involves the D-A-D (Decide-Announce-Defend) approach (Hunt, 2001). Nowadays such an approach is criticized for being more technocratic than democratic, and for being inefficient, socially unjust, and ethically biased. The Radioactive Waste Management Advisory Committee (2001) in the United Kingdom, for instance, has become aware of this and has declared such an approach "inappropriate".

Beyond the D-A-D approach there are a variety of ways to allow the public to enter into the decision-making process, some of which occur earlier or later along the decision chain with varying efforts to allow public input. For instance, Vári, Reagan-Cirincione, and Mumpower (1994) outline the four ways that public participation has been conceived and used by various nations as they strive to site nuclear facilities:

1. Stakeholders are involved in the project, receiving information, but have no decision-making power (equivalent to the D-A-D approach).
2. Stakeholders are granted the power to review and modify recommendations or decisions.
3. Stakeholders are given the power to make recommendations, although decision-making power is reserved by state or private agencies or institutions.
4. Stakeholders are given direct power to choose a solution or make a decision.

An example of the first way of involving the public, which had been used as the preferred public participation method in the United States up until recent times, is the public hearing. Many heavily criticize this form of public participation, however. For example, Kraft and Clary (1993) say that such participation provides a weak opportunity for real public involvement, lacks two-way channels of communication, and may be usurped by planners to promote the facility they are planning.

Evidence that suggests that maximum community involvement is more effective than minimal community involvement is offered by Carnes, Copenhaver, Sorensen, Soderstrom, Reed, Bjornstad, et al. (1983). They asked a sample of Wisconsin residents whether or not they would oppose a radioactive waste repository built in their state. Initially 26% of respondents indicated they would approve such a repository. Then after being offered accompanying conditions (which included independent monitoring, enhanced community control of the facility, and the power to shut the facility down) the percentage favoring the repository rose to 46%.

Whatever precise path of public participation that a mission-oriented organization might consider adopting, the Swiss-based Expert Group on Disposal Concepts for Radioactive Waste (EKRA), acknowledge there is always the question of how to reconcile different forms of knowledge, levels of rationality, and claims of truth and, at the same time, carry on pluralistic and democratic discussion on the topic of radioactive waste management (EKRA, 2000).

The public participation schemes undertaken by various authorities involved with nuclear waste policy and planning have brought to light a number of recurring public concerns. These are listed below:

A. What are the exclusion criteria for siting?

People want to know what reasons the authorized body would have to stop a siting (Wilson, 2000, p. 67).

B. Complaints of notification

Despite there sometimes being a lot of press, many people complain they don't know when or where the public consultation process is be to held, and that if they do know, they don't know what the parameters of the meeting are (Wilson, 2000, p. 39).

C. Statements of Uncertainty

People seem to want to have a clear statement of technical and scientific uncertainties up-front. In Canadian public hearings it was found that the public takes the uncertainties far more seriously than the experts, and trust is not built by scientific uncertainties not being stated up-front (Wilson, 2000, p. 37).

The public participation process in the Finnish case also found that there was some public unease about whether experts can claim certainty of their knowledge with regard to the long-term safety of the facility (Nuclear Energy Agency, Radioactive Waste Management Committee, 2002).

D. The right of veto

The right of veto is clearly desired (Flynn, Mertz and Slovic, 1993). But who should it be invested in—the citizenry, the local council, or low- or top-level Government officials? When in the process should it be given—before or after feasibility studies?

Douglas Easterling and Howard Kunreuther (1995, p. 12) offer a method of voluntarism in which a waste management organization proposes to prospective volunteer communities a list of minimum requirements for a facility. The waste managers can then ask the prospective volunteer communities to propose the conditions under which they would allow the facility to be constructed. Under Easterling and Kunreuther's vision of such a system, potential host communities enter into negotiations with the developer only if they are interested and that they can de-select themselves at a future time.

E. The reliance on experts

Nuclear experts are only occasionally seen as being neutral. Usually, the public discounts expert evidence because of whom the expert works for (Papinchak and Wingard, 1990). For example, "the international consensus on the concept [of deep geological disposal] comes from proponents of the industry only" (Wilson, 2000, p. 39).

The public often believes that experts have the prevalence of working towards:

- Making their employers happy (Johnson, 2003; Irwin, Dale and Smith, 1996).
- Justifying their own earlier judgments (Wynne, 1996; Sismendo, 1996).
- Legitimizing their own personal value framework (Sundqvist, 2002; National Research Council, Committee on Disposition of High-Level Radioactive Waste Through Geological Isolation, 2001; Slovic et al., 1993, p. 64).

Thomas Rosenberg of the Lovisa movement in Finland found that the EIA process at Eurojoki was steeped in scientific camouflage by the experts involved, alienating the citizenry from the decisions (Nuclear Energy Agency, Radioactive Waste Management Committee, 2002). According to a moderator within this process, some participants mentioned that due to lack of resources, some people who held suspicions about the proposed plant could not hire their own independent experts to offer independent views.

F. Measuring public attitudes

Lois Wilson (2000) asks, "What method will measure public acceptance? Referendum? Plebiscite? City Council vote?" (p. 43).

G. False participation

According to Wallentinus and Paivo (2001), there have been instances where bodies have sat down to listen to the stakeholders in a succession of meetings in which no effort was made to adjust proposals to the ongoing suggestions of stakeholders. This issue can be expanded to include the often expressed suspicion that much of what constitutes public participation is just public relations (Beder, 1999; Kraft and Clary, 1993), an attempt to manipulate public acceptance into a pre-chosen proposal (or to make the public choose from a range of favored proposals).

H. Trust and trustworthiness

Speakers at hearings and on citizen panels intimate that it is often very difficult to trust the various actors involved in radioactive waste management. For example, in European nations, Eurobarometer found that 29% of the respondents state that they are very worried about the way nuclear waste is handled in their own country and only 10% trust the information provided by the nuclear industry (INRA European Coordination Office, 2002).

Intergenerational Justice in Nuclear Waste Management

Issues of justice do not cut across only space but also across time. Radioactive waste is long-lived. The waste produced today is going to be around many years after this generation has disappeared. There are a number of ethical problems thrown up as a result of this and they tend to be categorized together in the literature under the rubric of *intergenerational equity*; a phrase meant to convey the fact that there are obligations and rights that the current generation owe to, or project upon, future generations.

Consent

In democratic societies it is often regarded as important to get the agreement of the local people in some way before building a hazardous facility. However, it is impossible to get the consent of future generations of communities that may surround such facilities. With regards to nuclear waste this becomes an intergenerational issue, for the waste remains hazardous from 100 to a million years. Even if the most extensive and intensive public participation, democratic decision-making and stakeholder involvement was all enacted, and even if local consent is given for a nuclear waste management facility to be constructed at this moment in time, this does nothing to allay concern that such processes and such consent decides the future environmental quality of peoples who have not, in any way, approved the facility. In light of this, Nilson (2001) raises a question: how far in the future can we make democratically credible decisions?

Relying on Future Techno-Fixes

While talking about our responsibility to manage our own radioactive waste, Shrader-Frechette (2000) makes the point that:

> Of course it may be counter-argued that future persons ought to bear more of the risk and cost of nuclear waste because those future people will be better prepared to face these technological and economic risks (p. 773).

Thus, given that society is always advancing and progressing scientifically, technologically and economically, radioactive waste managers shouldn't be in too much of a hurry to invent a solution because any generation that comes after us will provide a better one. Shrader-Frechette's answer to this is:

- We don't know what the future holds.
- Economically, demographically, and resource uncertainty may make it more difficult for future generations to solve these problems.
- Just because they *may* be able to solve the problem better than present generations, this does not mean they *should* solve our problems.

In contrast to Shrader-Frechette, Nilson (2001) sets store in the ability of each generation being able to furnish the next with the skills, resources and means to manage problems that they leave behind. She says that to do that we should make sure the chain of skills remains undiminished. To do that there needs to be a number of processes set in place (such as record maintenance, standardizing a long timescale review process, ensuring traditions of practice are sustained, etc.). Only by actively doing these things in the near-term can we possibly rely on social institutions that are supposed to preserve current experience and knowledge and pass it on to the future generations. The Nuclear Energy Agency (NEA) has also raised similar points (Nuclear Energy Agency, 1995). Critics of this approach, however, might raise the point that this is an apology for the continuation of a nuclear industry.

Social and Political Change

The social and political backgrounds against which radioactive wastes are to be managed are liable to change, perhaps drastically, in both the short-term and long-term life of the waste. Some, like Buser (1997), have noted that our knowledge of the physical environment and our prediction of its stability, while full of lacunae and doubts, are far more impressive than our ability to understand and predict the course of the social and political environment. "Political science fiction" is the phrase Lois Wilson (2000) is driven to use when cogitating about failing institutions and changing social circumstances over the lifetime of radioactive waste.

Writers like Wilson are sensitive to the fact that things are going to change quite unpredictably. It's not only the case that wars will be fought, economic slumps and booms will come and go, but that nations also may rise and fall. And even the concept of a nation may disappear (as some intimate with regard to globalization (Giddens, 2000)) taking along with it, perhaps, any institutional body charged with maintaining control or a watching brief over nuclear waste.

Given all of this, many have stated that now is the time to solve the problem, now is the time to think of a permanent solution (McCombie, Pentz, Kurzeme and Miller, 2000; McCombie and Chapman, 2002; Säteilyturvakeskus, 1989; Nuclear Energy Agency, 1995; International Nuclear Societies Council, 2002). Whether this is true or not, an important question that must come up is this: is it worthwhile making any predictions for the future of social environment as is done with the physical environment? Some, like Wilson (2000), would say no, since it is merely sooth-saying. The chances of you predicting the right result are very small. Others may say yes, but only if we acknowledge that our predictions are limited to generalities. It is possible, for example, for social scientists to arrive at a range of scenarios for future societies that are helpful in providing overall advice to today's radioactive waste managers. Given that most social science has never been a predictive art, except to those with a distinct utopian agenda, most sociologists would be skeptical of the social and political predictions. However, based on their attempts to delve into the social aspects of other environmental problems (Williams, 1998; Dunlap and Michelson, 2002), and based upon their attempts to tease out the social aspects within scientific and technological projects (Sismendo, 1996; Mack, 1990), most social scientists would be convinced of the massive importance of social and political issues on the future management of nuclear waste, and they'd probably say that these factors would equal or outweigh many of the technical factors already considered by nuclear waste managers.

Information Upkeep

If future generations are to be able to care for or avoid the radioactive waste facilities that this generation constructs, then some way of communicating the dangers of radioactive waste to these future generations has to be realized. However, any attempt to do this must be cognizant of the changing regimes of information storage. Mainstream manners of conveying information are obviously subject to change over long time periods. Many of the oral traditions and symbolic representations that were standard thousands of years ago are largely lost to or lost on the current generation. Similarly, the documents we produce now relating to the siting of dangerous waste are less likely to survive than the waste itself. The digital revolution may exacerbate this problem according to Ulrike Fink (1993) who points out that data losses may take place even faster due to the rapid progress and subsequent incompatibility of computer systems. Fink offers an example of information loss that she believes is somewhat analogous to what may happen with regards to nuclear waste:

> Everyone knows that Germans are especially tidy and painstaking—but nevertheless, now and then it happens that old pits of abandoned coal mines are just drilled by chance! That means, either the knowledge has got lost during 100 years or the people didn't study the available, existing data—the people were not conscious of the problem (p. 136).

This case, suggests Ulrike Fink, implies that despite our best record-keeping efforts, it is likely that the information we produce about our radioactive waste activities is probably going to be inaccessible to future generations.

References

Beder, S. (1999). "Public Participation or Public Relations?", in B. Martin (ed.), *Technology and Public Participation*. Wollongong, Australia: University of Wollongong.

Blowers, A. (1999). "Nuclear Waste and Landscapes of Risk", *Landscape Research*, 24(3): 241-263.

Blowers, A., Lowry, D. and Solomon, B.D. (1991). *The International Politics of Nuclear Waste*. London: Macmillan.

Brody, J.G. and Fleishman, J.K. (1993). "Sources of Public Concern about Nuclear Waste Disposal in Texas Agricultural Communities", in R.E. Dunlap, M.E. Kraft and E.A. Rosa (eds.), *Public Reactions to Nuclear Waste*. Durham, NC: Duke University Press.

Buckingham-Hatfield, S. (2000). *Gender and Environment*. London: Routledge.

Buclet, N. and Bouzidi, Y. (2003). "Confidence, Actor's Beliefs and Transparency: The Case of ANDRA", in K. Andersson (ed.), *Valdor 2003, Values in Decisions on Risk: Proceedings* (pp. 160-168). N.p.

Buser, M. (1997, August). "Guardianship versus Disposal: A Modern-Day Conflict with Implications for the Future", *Nagra Bulletin*, 30, 24-32.

Carnes, S.A., Copenhaver, E.D., Sorensen, J.H., Soderstrom, E.J., Reed, J.H, Bjornstad, D.J., et al. (1983). "Incentives and Nuclear Waste Siting: Prospects and Constraints", *Energy Systems and Policy*, 7(4): 324-351.

Dunion, K. (2003). *Troublemakers: The Struggle for Environmental Justice in Scotland*. Edinburgh: Edinburgh University Press.

Dunlap, R.E. and Michelson, W. (eds.) (2002). *Handbook of Environmental Sociology*. Westport CT: Greenwood Press.

Dunlap, R.E., Rosa, E., Baxter, R. and Mitchell, R. (1993). "Attitudes Toward Siting a High-Level Nuclear Waste Repository at Hanford, Washington", in R.E. Dunlap, M.E. Kraft and E.A. Rosa (eds.), *Public Reactions to Nuclear Waste*. Durham, NC: Duke University Press.

Easterling, D. and Kunreuther, H. (1993). "The Vulnerability of the Convention Industry to a High-Level Nuclear Waste Repository", in R.E Dunlap, M.E. Kraft and E.A. Rosa (eds.), *Public Reactions to Nuclear Waste*. Durham, NC: Duke University Press.

Easterling, D. and Kunreuther, H. (1995). *The Dilemma of Siting a High-Level Nuclear Waste Repository*. Boston: Kluwer.

Edelstein, M.R. (1988). *Contaminated Communities: The Social and Psychological Impacts of Residential Toxic Exposure*. Boulder, CO: Westview Press.

EKRA (2000). *Disposal Concepts for Radioactive Waste: Final Report*. Bern, Switzerland: Federal Office of Energy.

Everts, S.I. (1998). *Gender and Technology: Empowering Women, Engendering Development*. New York: St Martins Press.

Fahys, J. (2003, March 11). "Skull Valley Plan Rejected", *Salt Lake Tribune*.

Fink, U. (1993). "The Nuclear Guardianship: Concept for a Radioactive Future", in *Poison Fire, Sacred Earth: Testimonies, Lectures, Conclusions: The World Uranium Hearing, Salzburg 1992*. Munich: World Uranium Hearing, pp. 135-138. Available on the Web: http://www.ratical.org/radiation/WorldUraniumHearing/UlrikeFink.txt

Flynn, J.H., Slovic, P., Mertz, C.K. and Toma, J. (1990). *Evaluations of Yucca Mountain: Survey Findings About Attitudes, Opinions, and Evaluations of Nuclear Waste Disposal and Yucca Mountain, Nevada*. Carson City, NV: Nevada Nuclear Waste Project Office.

Flynn, J.H., Mertz, C.K. and Slovic, P. (1993). *The Spring 1993 Nevada State Telephone Survey: Key Findings*. Carson City, NV: Nevada Nuclear Waste Project Office.

Fowler, C.S., Hamby, M., Rusco, E.R. and Rusco, M.K. (1990). *Native Americans and Yucca Mountain: A Summary Report*. Carson City, NV: Nevada Nuclear Waste Project Office.

Gawande, K. and Jenkins Smith, H. (2001). "Nuclear Waste Transport and Residential Property Values: Estimating the Effects of Perceived Risks", *Journal of Environmental Economics and Management*, 42(2): 207-233.

Gerrard, M.B. (1996). *Whose Backyard, Whose Risk: Fear and Fairness in Toxic and Nuclear Waste Siting*. Cambridge, MA: MIT Press.

Giddens, A. (2000). *Runaway World: How Globalization is Reshaping Our Lives*. London: Routledge.

Gowda, M.V.R. and Easterling, D. (2000). "Voluntary Siting and Equity: The MRS Facility Experience in Native America", *Risk Analysis*, 20(6): 917-929.

Great Britain, Parliament, House of Lords, Select Committee on Science and Technology. (1999). *Management of Nuclear Waste*. London: Stationery Office.

Gregory, R.S. and Satterfield, T.A. (2002). "Beyond Perception: The Experience of Risk and Stigma in Community Contexts", *Risk Analysis*, 22(2): 347-358.

High Level Radioactive Waste Transportation Factsheet (2000). Las Vegas: Citizen Alert.

History and Some Facts to Wellenberg: Project of a Swiss LLW Repository (2002, September). Report Presented at the 3rd COWAM Seminar, Furigen, Switzerland.

Hornig Priest, S., Bonfadelli, H. and Rusanen, M. (2003). "The 'Trust Gap' Hypothesis: Predicting Support for Biotechnology Across National Cultures as a Function of Trust in Actors", *Risk Analysis*, 23(4): 751-780.

Hunt J. (2001). "Framing the Problem of Radioactive Waste: Public and Institutional Perspectives", in *Valdor 2001 Proceedings* (pp. 222-229). N.p.

INRA European Coordination Office (2002). *Eurobarometer 56.2: Europeans and Radioactive Waste*. N.p.: DG Press and Communication, European Commission.

International Atomic Energy Agency (2003). *The Long-term Storage of Radioactive Waste: Safety and Sustainability*. Vienna: International Atomic Energy Agency.

International Nuclear Societies Council (2002). *Current Issues in Nuclear Energy: Radioactive Waste*. La Grange Park, IL: American Nuclear Society.

Irwin, A., Dale, A. and Smith, D. (1996). "Science and Hell's Kitchen: The Local Understanding of Hazard Issues", in A. Irwin and B. Wynne (eds.), *Misunderstanding Science?: The Public Reconstruction of Science and Technology*. Cambridge: Cambridge University Press.

Johnson, B.B. (2003). "Further Notes on Public Response to Uncertainty in Risks and Science", *Risk Analysis,* 23(4): 741-789.

Kleindorfer, P.R., et al. (1988). *Valuation and Equity in the Siting of Nuclear Waste Repository*. Carson City, NV: Nevada Nuclear Waste Project Office.

Kraft, M.E. and Clary, B.B. (1993). "Public Testimony in Nuclear Waste Repository Hearings: A Content Analysis", in R.E Dunlap, M.E. Kraft and E.A. Rosa (eds.), *Public Reactions to Nuclear Waste*. Durham, NC: Duke University Press.

Kraft, M.E., Rosa, E.A. and Dunlap, R.E. (1993). "Public Opinion and Nuclear Waste Policymaking", in R.E. Dunlap, M.E. Kraft and E.A. Rosa (eds.), *Public Reactions to Nuclear Waste*. Durham, NC: Duke University Press.

Krannich, R.S., Little, R.L. and Cramer, L.A. (1993). "Rural Community Residents' Views of Nuclear Waste Repository Siting in Nevada", in R.E Dunlap, M.E. Kraft and E.A. Rosa (eds.), *Public Reactions to Nuclear Waste*. Durham, NC: Duke University Press, pp. 263-287.

Mack, P.E. (1990). *Viewing the Earth: The Social Construction of the Landsat Satellite System*. Cambridge, MA: MIT Press.

Mariotte, M. (1998). *Radioactive Waste Transport: The German Experience*. Retrieved from Nuclear Information and Resource Service Web site: http://www.nirs.org/factsheets/radwastegermany.htm

Martin, B. (ed.) (1996). *Confronting the Experts*. Albany, NY: State University of New York Press.

McAvoy, G.E. (1999). *Controlling Technocracy: Citizen Rationality and the NIMBY Syndrome*. Washington, DC: Georgetown University Press.

McCombie, C. and Chapman, N. (2002). "Sharing the Waste Burden", *Nuclear Engineering International,* 47(580): 27-30.

McCombie, C., Pentz, D.L., Kurzeme, M. and Miller, I. (2000). "Deep Geological Repositories: A Safe and Secure Solution to Disposal of

Nuclear Wastes", in *GeoEng 2000: An International Conference on Geotechnical & Geological Engineering, 19-24 November 2000, Melbourne, Australia*. Lancaster, PA: Technomic.

National Research Council, Committee on Disposition of High-Level Radioactive Waste Through Geological Isolation (2001). *Disposition of High-Level Waste and Spent Nuclear Fuel: The Continuing Societal and Technical Challenges*. Washington, DC: National Academy Press.

Nevada Nuclear Waste Project Office (1999). *Transportation of Spent Nuclear Fuel and High-Level Radioactive Waste to a Repository: Factsheet*. Retrieved from http://www.state.nv.us/nucwaste/trans/trfact03.htm

Nuclear Energy Agency (1995). *The Environmental and Ethical Basis of Geological Disposal of Long-Lived Radioactive Wastes*. Paris: OECD.

Nilson, A. (2001). *Responsibility, Equity and Credibility-Ethical Dilemmas Relating to Nuclear Waste*. Stockholm, Sweden: Komentus.

Nuclear Energy Agency, Radioactive Waste Management Committee (2002). *Forum on Stakeholder Confidence, 2nd FSC Workshop: Executive Summary and International Perspective, Stakeholder Involvement and Confidence in the Process of Decision-Making for the Disposal of Spent Nuclear Fuel in Finland, 15-16 November 2001, Turku, Finland*. N.p.: OECD. Retrieved from http://www.nea.fr/html/rwm/docs/2002/rwm-fsc2002-1.pdf

Nuclear Energy Agency, Radioactive Waste Management Committee (2003). *Forum on Stakeholder Confidence: Canadian Site Visit and Workshop: Summary and International Perspective*. N.p.: OECD. Retrieved from http://www.nea.fr/html/rwm/docs/2003/rwm_fsc_2003_8.pdf

Nuclear Guardianship Project (1994). *An Ethic of Nuclear Guardianship: Values to Guide Decision-Making on the Management of Radioactive Materials*. Retrieved from http://www.nonukes.org/r02ethic.htm

Nuclear Shipping Accidents: Rare but Regular (2002). Retrieved from the Radio-Television News Directors Association & Foundation Web site: http://www.rtnda.org/resources/tipsheet/oct16.shtml

Oser, W. and Brown, M.Y. (1996). "A Background Briefing on Radioactive Pollution", Retrieved from http://www.nonukes.org/briefing.htm. Originally Published as: The Guardianship Ethic to Radioactive Pollution: A Meta Model for a Sustainable World. *Current World Leaders International Issues,* 38(6).

Oughton, D.H., Bay, I., Forsberg, E-M., Hunt, J., Kaiser, M. and Littlewood, D. (2003). *Social and Ethical Aspects of Countermeasure*

Evaluation and Selection: Using an Ethical Matrix in Participatory Decision-Making. Retrieved from http://www.strategy-ec.org.uk/output/reports/del4%20final%20version.pdf.

Papinchak, S. and Wingard, L. (1990, October 21). "Poll Shows Strong Opposition to Dump", *Las Vegas Review Journal*, 10A-12A.

Posiva Oy. (1999). *The Final Disposal Facility for Spent Nuclear Fuel: Environmental Impact Assessment Report*. Helsinki, Finland: Posiva Oy.

Rabe, B.G. (1994) *Beyond NIMBY: Hazardous Waste Siting in Canada and the United States*. Washington, DC: Brookings Institute.

Radioactive Waste Management Advisory Committee (2001). *Advice to Ministers on the Process for Formulation of Future Policy for the Long-Term Management of UK Solid Radioactive Waste. London: Dept. for Environment, Food and Rural Affairs.*

Rosa, E.A., Dunlap, R.E. and Kraft, M.E. (1993). "Prospects for Public Acceptance of a High-Level Nuclear Waste Repository in the United States: Summary and Implications", in R.E. Dunlap, M.E. Kraft and E.A. Rosa (eds.), *Public Reactions to Nuclear Waste*. Durham, NC: Duke University Press, pp. 291-328.

Säteilyturvakeskus (1989). *Disposal of High Level Radioactive Waste: Consideration of Some Basic Criteria*. Helsinki, Finland: Finnish Centre for Radiation and Nuclear Safety.

Shrader-Frechette, K. (1991). "Ethical Dilemmas and Radioactive Waste: A Survey of Issues", *Environmental Ethics,* 13: 327-343.

Shrader-Frechette, K. (1993). "Burying Uncertainty: Risk and the Case Against Geological Disposal of Waste", Berkeley: University of California Press, Berkeley.

Schrader-Frechette, K. (2000). "Duties to Future Generations, Proxy Consent, Intra- and Intergenerational Equity: The Case of Nuclear Waste", *Risk Analysis,* 20(6): 771-778.

Shrader-Frechette, K. (2001). "Risky Business: Nuclear Workers, Ethics and the Market Efficiency Arguments", *Ethics and the Environment,* 7: 1-19.

Sismendo, S. (1996). *Science Without Myth: On Constructions, Reality and Social Knowledge and Society*. Albany, NY: State University of New York Press.

Sjölander, A. (2003). "The Nuclear Waste Discourse in Malå: A Truth-Seeking Practice", in K. Andersson (ed.), *Valdor 2003, Values in Decisions on Risk: Proceedings* (p. 169). N.p.

Slovic, P., Layman, M. and Flynn, James. (1993). "Perceived Risk, Trust and Nuclear Waste: Lessons from Yucca Mountain", in R.E.

Dunlap, M.E. Kraft and E.A. Rosa (eds.), *Public Reactions to Nuclear Waste*. Durham, NC: Duke University Press.

Slovic, P. and Flynn, J.H. (1991). "Perceived Risk, Trust and the Politics of Nuclear Waste", *Science,* 254(5038): 1603-1607.

Sundqvist, G. (2002). *The Bedrock of Opinion: Science, Technology and Society on the Siting of High-Level Nuclear Waste*. Dordrecht: Kluwer.

Vári, A., Reagan-Cirincione, P. and Mumpower, J. (1994). *LLRW Disposal Facility Siting: Successes and Failures in Six Countries.* Dordrecht: Kluwer.

Wallentinus, H-G. and Paivo, J. (2001). "The Hallandsas Railway Tunnel Project in Sweden", in T. Hilding-Rydevik (ed.), *The Role of Environmental Impact Assessment in the Planning and Decision Process of Large Development Projects in Nordic Countries.* Stockholm, Sweden: Norderegio Report.

Wigley, D. and Shrader-Frechette, K. (eds.). (1994). *Comments on the Draft Environmental Impact Statement for the Construction and Operation of Clairborne Enrichment Center, Homer, Louisiana.* Vol. 2 of *Final Environmental Impact Statement for the Construction and Operation of Clairborne Enrichment Center, Homer, Louisiana.* Washington, DC: U.S. Nuclear Regulatory Commission.

Wile, R. and Cox, J.R. (2002). *What If: Nuclear Waste Accident Scenarios in the United States*. Environmental Working Group.

Williams, J. (1998). "Knowledge, Consequences and Experience: The social Construction of Environmental Problems", *Sociological Inquiry*, 68(4): 476-497.

Wilson, L.M. (2000). *Nuclear Waste: Exploring the Ethical Dilemmas*, Toronto, Ontario: United Church Publishing House.

Wynne, B. (1996). "Misunderstood Misunderstandings: Social Identities and Public Uptake of Science", in A. Irwin and B. Wynne (eds.), *Misunderstanding Science: The Public Reconstruction of Science and Technology*. Cambridge, MA: Cambridge University Press.

7

Integrated Coastal Resource Management: A Prescription for Sustainable Development

Brian J. English

Environmental problems due to the human impact of growing urban populations in coastal areas necessitate an integrated approach to coastal resource management as a move toward sustainable development. Steps toward sustainable development of coastal communities should involve researchers, educators, and planners from a coordinated network of international aid agencies, non-governmental organizations (NGOs), national policy makers, and local communities. Hence, this paper aims to explain how international aid agencies and NGOs interact with national policy makers and local communities in integrated coastal management programs to promote sustainable development. From an examination of international agencies' roles in development, NGOs emerge as a counterweight to balance the perspective of development. Specific examples of several integrated coastal management programs explicate an understanding of organizational and institutional interaction in this process. Finally, some comments about social movements and their appropriateness for raising consciousness about environmental degradation through non-formal education programs will be given.

There is nothing more difficult to carry out, nor more doubtful of success, nor more dangerous to handle, than to initiate a new order of things.

Machiavelli, *The Prince*, 1513

Throughout history civilizations have developed on the water's edge. For thousands of years humans have had a close relationship with the sea. The vast resources of the oceans are an essential element for the survival of coastal communities. Ironically, the world's marine environments are also disposal systems for human generated waste. It may be an overgeneralization, but there is some truth in saying, "Everything comes from the sea and everything returns to the sea." The implication is that since the oceans have limited resources, there is a need to minimize the impact that the growing size and number of coastal communities have on marine environments. Although non-coastal communities also have pressing environmental issues, maintaining sustainability in coastal development is particularly important because "more than half the world's population lives within 60 km of the shoreline, and this could rise to three quarters by the year 2020" (United Nations Conference on Environment and Development [UNCED], 1992, para. 7. 3). At the advent of the 21st century there are 20 cities in the world with a population of over 10 million. Sixteen of those cities are along coastlines. Growing populations not only put a greater strain on already depleting resources, but also are responsible for the degradation of marine habitats for those resources. According to the report of the Independent World Commission on the Oceans (1998), over 70% of the world's fish stocks are being exploited at or even beyond sustainable limits. "Over 80% of all marine pollution originates from land-based sources which are primarily industrial, agricultural and urban" (UNCED, 1992). The need to rethink development planning in coastal areas is undeniable. Steps toward sustainable development of coastal communities should involve researchers, educators, and planners from a coordinated network of international aid agencies, non-governmental organizations (NGOs), national policy makers, and local communities. Hence, this paper aims to explain how international aid agencies and NGOs interact with national policy makers and local communities in integrated coastal management programs to promote sustainable development.

After defining the problem of development in coastal areas this paper will briefly discuss sustainable development as a goal, and integrated coastal management (ICM) as a vehicle for moving toward that goal. From an examination of international agencies' roles in development, NGOs emerge as a counterweight to balance the perspective of development. Specific examples of several integrated coastal management programs explicate an understanding of organizational and institutional interaction in this process. Finally, some comments about social movements and their appropriateness for raising consciousness about environmental degradation will be given.

Growing Anthropogenic Pressures on Coastal Environments

The growing world population, world economy, and world trade have increased the demand for marine and coastal resources. Technology has enabled large companies to widen the range of resources they are able to exploit. Technology has also allowed an exponential increase in the "development" of coastal areas. Among the practices that contribute to unsustainable development are logging in mangrove forests, mining, dredging, filling, channelizing wetlands, bottom trawling, and dynamite fishing on coral reefs. Additional damage to marine ecosystems comes from land-based activities such as siltation from logging, fertilizer and pesticide run-off from agriculture, sewage effluent, industrial discharges, road building, urbanization of shorelines, construction of sea walls and jetties, and construction of harbors and marinas. The building of dams alters the flow of freshwater and nutrients into marine systems and prevents some aquatic species from reaching their spawning grounds (UNCED, 1992).

Some of the inherent implications of unsustainable development are summarized in *The Ocean, Our Future,* a report of the Independent World Commission on the Oceans (1998):

> All too often the costs [of unsustainable development] bear most heavily on the poor. Management regimes to control land-based sources of pollution, over fishing, and other threats to the stability of marine and coastal eco-systems are often poorly developed or do not exist. Access to many

> ocean resources is unregulated and, where such regulation does exist, it is often ineffective (p. 97).

Without a system for managing the ocean's resources, development will go unchecked. In market economies it is natural for individuals and companies to make decisions in their own financial interests. Unfortunately, short-term gains usually overshadow long-term consequences. From the types of detrimental practices listed above, it can be inferred that many decisions are being made without considering the long-term social and economic costs of degrading the environment. This has a tendency to worsen the cycle of poverty. The people profiting from the use of those resources are able to ignore the costs they impose on others:

> Liberalization may have yielded many economic benefits, but it has also increased the risk of environmental damage due to the failure of markets to register the environmental costs of economic activity. Markets have typically failed to signal the true scarcity of resources, both on- and offshore, and liberalization has often weakened corrective regulations or incentives. Indeed, the worldwide structure of property rights, taxes and subsides has encouraged over-use of coastal and marine resources. This has often placed these resources under intolerable stress (Independent World Commission on the Oceans, 1998; p. 98).

In their analysis of the World Bank's case studies on linkages between economy-wide policies and the environment, Cruz and Munasinghe (1996) further explain the paradoxical relationship between sustainable development and liberalization:

> Environmentally detrimental policy distortions include export promotion and trade liberalization that increase the profitability of exporting resources, particularly if those resources are underpriced or subsidized. Market failures occur when prices do not adequately reflect external effects such as pollution. Institutional constraints include poor accountability in state-owned enterprises, inadequately defined property rights, and weak financial intermediation,

all of which lower the incentives for sustainable resource management (p. 7).

As resources reach economic depletion, workers are forced out of jobs. "In the United States alone, coastal areas provide 28 million jobs—many of which will cease to exist if fisheries are contaminated, tourist destinations fouled or public health problems exacerbated" (Claussen, 1997). Through the loss of jobs and the depletion of economic resources in less developed countries, environmental degradation can lead to an increase in environmental refugees.

Although lowered incomes and unemployment affect both men and women, it would seem that such sociological problems harm women more. According to the *UN Chronicle*, "women are among those who suffer most from environmental degradation and also among the most significant actors in the conservation and safeguarding of natural resources" (Empowering Women, 1995, p. 47). As jobs become scarce, women are the first to be pushed out of the market. Some may leave rural communities for jobs in already overcrowded cities. The lack of skills, inadequate education, or limitations of only speaking a provincial language is likely to make it difficult for some to find jobs in the cities; thus forcing young women and even children into prostitution. Less money often means less food in developing countries. Women and children are the most likely to suffer from lack of nutrition. Lack of nutrition is often linked to high infant mortality, disease, and problems during pregnancy (Warren, 1996).

Sustainable Development

Sustainable development should not be interpreted as a fixed notion, but should rather be seen as "a process of change in the relationships between social, economic, and natural systems and processes. These interrelationships present a challenge to us in reconciling economic and social progress with safeguarding the global life support systems" (van Ginkel, 1998, Summary section, para. 2). Although sustainable development has many definitions, the most widely used definition comes from the World Commission on Environment and Development in its 1987 report *Our Common Future*, "Sustainable development is development that meets the needs of the present without compromising the ability of future generations to meet their own needs" (p. 43). It contains within it two key concepts:

- the concept of *needs*, in particular the essential needs of the world's poor, to which overriding priority should be given; and
- the idea of limitations imposed by the state of technology and social organization on the environment's ability to meet present and future needs.

It is conceivable that this definition has different meanings for different groups. Much of the rhetoric in *Agenda 21* and similar documents have tones that sustainable development is globally-managed interdependence. This interdependence has international development agencies defining appropriate development strategies for regions in the south, while regions in the north benefit from the south's sale of cheaper goods and natural resources. The critics of this view would argue that such economically driven globalization is not compatible with sustainable development because multi-national corporations and elite groups are more concerned with increasing their own profits than establishing a stewardship for the environment. Unfortunately, it is those corporations and elite groups that have monetary, political, and social power to influence national and international decisions on environmental policy. Therefore, sustainable development should come from a grassroots, community-based, self-reliant approach. An integrated approach of both top-down and bottom-up strategies is necessary to maintain a balance that promotes monitored national growth while minimizing local environmental impact.

Integrated Coastal Management

The concept of integrated coastal management (ICM) has evolved out of the United States' extensive coastal management practices that began in the early 1970s (Thai-Eng, 1998). "Integrated coastal management is a continuous, iterative, adaptive, and consensus-building process comprised of a set of related tasks, all of which must be carried out to achieve a set of goals for the sustainable use of coastal areas" (Bower, Ehler, and Basta, 1994). "Integrated" is the definitive word in integrated coastal management. The term has five separate dimensions (National Oceanic and Atmospheric Administration, 2000). The first dimension, intersectoral integration, is integration among different coastal and marine sectors including oil and gas development, fisheries, coastal tourism, marine mammal protection and port development. It is also integration between marine sectors and land-based sectors that affect the coastal and ocean environment, such as agriculture, forestry and mining. Intersectoral integration also implies a consciousness about and desire to resolve conflicts among government agencies in different sectors. This particular type of integration is essential in the planning of effective coastal management. It promotes a sharing of knowledge and commitment among actors.

A second dimension of the word integration is intergovernmental integration. This type of integration helps to realize the variance in perspectives of different levels of government so as to facilitate harmony among national, provincial and local governments in addressing public needs. This harmony should help to coordinate development policies. A third dimension, spatial integration, further connects land-based activities with coastal impacts.

The fourth dimension of integration is science management integration or the vital sharing of knowledge and information among the various disciplines concerned with coastal resource management. It also comprises the sharing of information and data with other actors in the resource management process for purposes of monitoring changes in the environment and revising coastal development planning. Finally, the fifth dimension of integration poses the greatest challenge. International integration focuses on minimizing and quelling international disputes over fishing activities, transboundary pollution, establishment of maritime boundaries, passage of ships and other issues.

The Role of International Development Agencies and Institutions

Beginning in the last quarter of the 20th century, there has been a significant increase in the international community's involvement in environmental issues, coastal development, and use of the ocean's resources. Perhaps the most publicized assertion of commitment comes from the United Nations Conference on Environment and Development's *Agenda 21*, chapter 17, that outlines international guidelines for the protection of the oceans. This document includes extensive information on human impact to the environment, plausible measures to lessen that impact, and warnings about possible environmental threats to human such as global warming and rising sea levels.

International agencies like the United Nations Development Program (UNDP), United Nations Environmental Programme (UNEP), UNESCO, NOAA, Food and Agriculture Organization of the United Nations (FAO), and the World Bank are significant actors in the integrated coastal management process. In spite of their pitfalls, they supply undeniably necessary resources to the process. UNEP functions include environmental impact assessments, promotion of regional cooperation, providing technical advice to governments in collaboration with UNDP and the coordination of scientific research. The information technology available to these international agencies is needed for accurate assessment and monitoring of environmental changes. These agencies also have financial capital or the networking means of coordinating the capital for development projects. Another important role of international governing institutions is their ability to create legally binding instruments to phase out and eliminate persistent organic pollutants and restrict inappropriate practices in the exploitation of ocean resources (Claussen, 1997). These types of agreements must be negotiated multilaterally, but can address land-based marine pollution and other problems.

The report of the Independent World Commission on the Oceans (1998) lists the positive developments that have come from international governance of the oceans. Several of these are:

- Governments have begun to call on civil society in attempts to reconcile conflicts in the oceans. This is encouraging greater awareness within society of the role of stakeholders in ocean policy and of the role of both national and international arrangements in articulating their interests in achieving policy objectives.

- There are legal foundations for the promotion of greater equity in the oceans, notably the concept of *common heritage of mankind*.
- The combination of terrestrial and marine uses in the coastal zone has focused new attention on integrated coastal management, and has compelled national and international bodies to assess their disparate approaches and guidelines.
- There is growing recognition of the need to think in terms of systems that encompass both sea water and fresh water. Recognition of the interdependencies existing between fresh water drainage basins, wetlands and the coastal zone is giving rise to the new international initiatives in integrated watershed management which take into account impacts on coastal and marine areas.
- The ecosystem-based and precautionary approaches to the management of living resources have spread from the regional level to the global level, reflected in the 1995 Agreement on Straddling Fish Stocks and Highly Migratory Fish Stocks.
- At a policy level, the linkages between international legal arrangements for the oceans, biodiversity, wetlands, rivers, the atmosphere, movements of hazardous wastes, and other related issues have been explicitly incorporated into some regional arrangements, notably for the Mediterranean.
- Of particular importance is the Global Programme of Action for the Protection of the Marine Environment from Land-Based Activities, adopted in Washington in 1995, which has established a clearinghouse mechanism under UNEP auspices, with the prospect of

linkages to the marine and coastal biodiversity issues (pp. 141-142).

Although there have been positive developments from international agencies being involved in the governance of the oceans, there are many criticisms about their effectiveness in establishing sustainable development. What is immediately noticeable in the above list is the lack of mention of any improvement to the quality of life as a result of improved environmental conditions for those living in the south. The effectiveness of high-tech monitoring systems and heightened international awareness of environmental degradation is only meaningful if the outcome is improved social conditions as a bi-product of improved environmental conditions. Some critics are very skeptical about the rhetoric in *Agenda 21*. Imber (1994) aptly summarizes this criticism:

> For sustainable development to represent a cultural-shift rather than an additional slogan, the UN system will require more than the paper-commitments of Agenda 21. It will require an overhaul of the mechanisms for coordination, and massive net additional funding. The lesson of the last decade is that whereas finance is definitely not forthcoming without structural reform, such reform is unlikely (p. 130).

Top-down strategists may argue that this additional funding would come from economic expansion. This is evident in the words of the World Bank (1992): "Rising incomes with sound environmental policies and institutions can form the basis for tackling both environmental and development problems. The key to growing sustainable is not to produce less but to produce differently" (p. 36).

However, without clear definitions of what the World Bank means by "differently", the paradox of the liberal-institutionalist approach to sustainable development remains. Williams (1996) claims that these liberal ideologies of institutions, such as the World Bank, regard environmental degradation as an accidental outcome of development and that the environmental problems can be easily rectified through market-led solutions. It is illogical to believe that a solution will come from the same brand of capitalism that led to exponential increases in environmental degradation since the onset of the industrial revolution. Therefore, a counterweight to the power and influence of international institutions and agencies is needed to balance the perspective on development. This counterweight is the grassroots, bottom-up movement widely supported by NGOs.

The Role of Non-Governmental Organizations

Fernandes (1985) aptly notes, "the function of non-governmental organizations is to serve" (p. 7). Non-governmental organizations (NGOs) offer a *bottom-up* inductive approach to bettering communities and addressing human concerns. Rubem discusses how NGOs in Latin America have promoted people's participation in slum management and encouraged political involvement. In a specific report on NGOs in the field of education, Archer (1994) reviews how some NGOs have historically focused on providing educational services to communities throughout the world. Although it is certain that many of these organizations serve causes not directly related to environmental issues, some of the other important literature on NGOs makes direct references to their involvement in environmental programs. In an overview of environmental politics in Asia, Schubert (1993) affirms that in most Asian nations, NGOs are "the primary impetus for environmental protection and nature conservation" (p. 241). According to Schubert, many of the thousands of environmental NGOs in Asia are "grass roots movements of people concerned about specific conditions in local eco-systems" (p. 241).

Whether NGOs focus on environmental issues or seek to provide other services, they provide additional monetary backing for local community efforts. Fernandes' (1985) claim that there are thousands of NGOs administering hundreds of millions of dollars in Latin America (p. 7) gives an example of how NGOs can provide needed resources for grass roots efforts. Schubert (1993) reasons that despite good intentions, many governments lack sufficient funding, training, and enforcement to implement effective environmental protection policies and programs. Therefore, there is a need for NGOs to augment the environmental efforts of national governments. Schubert clearly states, "The insufficiency of resources available to most policy makers in Asian nations calls for, even necessitates, the active inclusion of NGOs in policy formulation, enactment and enforcement" (p. 242). This is one way of integrating the actors involved in coastal management programs.

Although Schubert is referring to a broad view of NGO roles in developing effective environmental policies, there are definite implications for NGO roles in more specific environmental education programs. NGOs can provide resources to increase the probability that community-based environmental programs will be effective agents of change. Ideally, educational programs can promote attitudinal and behavioral changes to facilitate policy formulation, instigate action and reduce the burden of enforcement.

Invasive or Noninvasive?

Fernandes (1985) provides a list of organizational problems that many NGOs have. These include, inconsistencies between goals and budgets, poorly paid staff, and inter-organizational communication problems. Proposing solutions to these problems is beyond the scope of this paper; however, it is necessary to address criticisms of NGOs that are relevant to the theoretical design of eco-centered community-based environmental education.

Some critics of NGO involvement in local community concerns may argue that many NGOs are actually products of governments that are set up to implement official agenda (Quizon and Reyes as cited in Toh and Floresca-Cawagas, 1997). The argument would be that governments use NGOs to disguise political agenda. Similarly, Toh and Floresca-Cawagas argue that, "there are differences in world views and motivations among NGOs, some of which may not be authentically dedicated to the well-being of their constituents" (p. 534). If NGOs have goals that do not address the real needs of the community it is likely that their involvement will be seen as an outsider attempt to control local social institutions. When locals view NGOs as outsiders, resentment will grow and participation will wane.

It is important that NGOs avoid becoming invasive in their involvement. This is especially true for NGO support in establishing community-based educational programs in rural areas. In addition to the theoretical reasons already discussed in this paper, there are the practical reasons of workability for NGOs to maintain a non-invasive approach to implementing educational programs. In reference to non-formal education, van Riezen (1996) reasons that since specific groups have specific needs, the curriculum used in an educational program must relate to the needs and resources of the local community. People in rural communities will not profit from curriculum and textbooks designed for people in cities or more affluent countries. K. van Riezen explains that avoiding invasive involvement includes using the vernacular as a way of showing respect for the local culture. The inclusion of local culture sends a message to communities that their participation is valued. This gives community members a sense of worth and purpose that encourages active participation.

McCormick (1993) gives two factors that influence the effectiveness of NGOs. These, too, apply to NGOs in general, but also have significant relevance to community-based educational programs. The two factors are:

- Their political influence (as measured by the level of political support they enjoy, and their ability to use political structures effectively); and
- The importance of having clearly defined constituencies and clearly defined avenues through which to make their appeals and to influence government (p. 142).

Although McCormick contends that NGOs need clearly defined constituencies and strategies to be effective, others may reason that this is not necessary. The famous Chipko movement in India is an example of an effective campaign that grew to a critical mass while remaining loosely organized (Viswanath, 1991). The Chipko movement is a case where locals' values and love for their forest homeland motivated them to unite in an activist campaign against foreign logging companies. This campaign began without clearly defined avenues through which to make their appeals or to influence the government. Although the group of activists better fits the definition of a movement than an NGO, it exemplifies one type of local organization in developing countries. Viswanath more clearly defines the range of local NGOs in India as being a "mixed bag, ranging from service-oriented groups to militant movements, usually of the left" (p. 37). The Chipko is an example of one community organizing to campaign for a local cause but having far reaching influence in motivating other communities to form grass roots initiatives.

In their discussion on people-centered education in the Philippines, Toh and Floresc-Cawagas (1997) suggest four themes. The four themes are: a pedagogy of dialogue; a praxis of critical empowerment; active nonviolence for peace and justice; and walking in solidarity. Consideration of these four themes allows community educators and NGOs to negotiate how to maximize the positive impacts of McCormick's two factors. The first theme can begin to clearly define participants, methods and approaches to achieving the goals of a community-based education program. As previously discussed, empowerment, the second theme, can lead to political clout beyond the local level. Thus, the first two themes address McCormick's second conditional factor for effectiveness. The second, third and fourth themes are all significant in optimizing the political influence that an NGO backed grassroots environmental program might have.

It is essential to carefully weave all four of the themes in the planning and implementing of community-based coastal resource management programs. Inclusion of these themes will help to strengthen ties between any supporting NGO and the coastal community. Inclusion of these themes will help to ensure a greater effectiveness in achieving the goals of a community-based environmental program.

Examples of Integrated Coastal Management

SEACAM

One of the newest examples of the implementation of an ICM program is in East Africa. In 1997, the Secretariat for East

African Coastal Area Management (SEACAM) was formally established to work with stakeholders in 10 coastal countries; Comoros, Eritrea, Kenya, Madagascar, Mauritius, Mozambique, Reunion, Seychelles, South Africa and Tanzania. The objectives of the SEACAM project are to assist the Eastern African coastal countries in implementing and coordinating coastal management activities in the region. After one year, Voabil, Engdahl and Banze (1998) conducted an evaluation of the project. This section summarizes their findings.

In the first year SEACAM was able to implement the most intensive effort in coastal management training ever held in Eastern Africa. The authors claim that almost 80 stakeholders benefited from SEACAM's courses in Project Development and Management, and Environmental Assessment of Coastal Tourism. However, the authors do not specify who the 80 stakeholders were or to what degree they benefited.

A major priority in the first year of the program was the capacity building of local NGOs. This includes a series of five week-long training courses throughout the region. These courses are based on SEACAM's training manual. This program was able to reach over 60 NGOs in six of the countries. The authors report evidence that these programs help build confidence for designing projects among NGO representatives. Self-designed projects alleviate the need for outside consultants.

SEACAM has elicited the advice of other coastal management programs so that the projects in Eastern Africa can learn from the successes and failures of existing activities. SEACAM brought together 66 coastal management practitioners from 18 countries around the world. These practitioners represented research organizations, national and international NGOs, donor agencies and national governments. Representatives from the World Bank were present at four of the courses. This is an important component for newly established programs. Shared experiential learning from other organizations can save time and money, thus increasing effectiveness.

Since the project was only a year old at the time of the report, the focus was on the success of training and dissemination of information. The authors concluded that the efforts in these areas were successful and with sustainable financing should grow to be even more effective. From their report, it is evident that this is a major first step toward integrated coastal management in Eastern Africa. However, a longer case study at specific sites is needed to determine the true effectiveness of the ICM process in this region. The following section summaries a report on a group of case studies conducted in Asia.

Integrated Coastal Management in Southeast Asia

Chua Thai-Eng (1998) assessed the performance of eight ICM projects in Southeast Asia. These projects were developed through the ASEAN/US Coastal Resource Management Project and the Global Environment Facility (GEF)/UNDP/IMO Regional Programme for the Prevention and Management of Marine Pollution in the East Asian Seas. These section summaries the author's discussion of lessons learned during the 13 years of practicing ICM in seven countries; Singapore, Malaysia, Thailand, China, the Philippines, Brunei and Indonesia. This first ICM was launched in Southeast Asia by the United States Agency for International Development (USAID). An international NGO, the International Center for Living Aquatic Resource Management, was in-charge of implementing the program from 1986 to 1992. Since 1994, the GEF and UNDP have been involved in ICM programs in Southeast Asia. The International Maritime Organization (IMO) is in-charge of executing these programs. Swedish International Development Agency (SIDA) and Science of Anticipation*Recognition*Evaluation*Control of Health Risks (SAREC) are working collaboratively in the training process. One advantage of this collaboration is that it leads to closer south-south cooperation through staff exchange and training workshops. It also helps to strengthen institutional capacity in Southeast Asian developing nations such as Vietnam. This type of exchange helps to take the place of hiring foreign experts (Hallak, 1990).

The author found that in most of the project sites, people's organizations and NGOs were less significant than other actors because stakeholders were confined mostly in government agencies and local authorities. Still, there was a significant increase in public awareness of environmental degradation at all sites. The two most successful sites were Batangas, Philippines, and Xiamen, China. Both sites have benefited from the experiences and lessons learned in other ICM initiatives around the world. At these two sites, perception and attitude changes lead to stronger political commitments and allocation of financial resources for environmental projects. In Xiamen there was definite evidence of improved environmental conditions, a first step toward sustainable development. The Yaundang lagoon clean up in Xiamen resulted in improved water quality through management interventions, and use conflicts were minimized. Similar results in Batangas prompted the government to establish the Provincial Government-Environment and Natural Resource Office. The author notes that an important development of the ICM process in Batangas was the change in perception of industries' relationship with the environment. "Perception changes occur when people are convinced of the values and significance of ICM initiatives" (Hallak, 1990, p. 605). Creating public awareness among stakeholders should be a continual process. Public awareness helps to generate support, raise consciousness about environmental issues and promote civil advocacy for environmental protection and sustainable development. Another positive aspect of the ICM process was the development of a critical mass of technical experts from participating countries.

Outside of creating awareness and developing ICM projects, there was little evidence of impact at the other six sites. However, there have been several successful initiatives for mangrove reforestation at several sites.

Despite the lack of technical know-how and experience of local governments, their participation is essential. This is evidenced in the comparative successes of the Xiamen and Batangas initiatives, where local government ownership of the ICM programs was apparent. In the other sites, local governments were only minimally involved. The author stresses the importance of stakeholder's participation in ICM:

> The stakehoders denotes all sectors of society at the local level that are directly or indirectly affected by exploitation

and use of coastal resources. They include sectors that exploit or use the natural resources for profit, coastal communities that traditionally use natural resources for their food and livelihood, and the public sectors (local and central) that govern and manage the use of these resources (Hallak, 1990, p. 606).

The author also warns against an ICM initiative trying to solve too many problems at the same time because many of the issues that are related to political, cultural, or socioeconomic systems can take a long time to work through. Therefore, ICM initiatives should not try "quick fixes" and expect miracles.

Conclusion

This paper has attempted to demonstrate how an integrated top-down and bottom-up approach to coastal management is necessary to provide a balanced perspective of sustainable development. International governing institutions, aid agencies, and local communities differ not only in their perspectives, but also in their strengths and weaknesses. Therefore, integration of these actors is quintessential in the continual strife toward sustainable development. This integration should focus on the strengths of each, while seeking ways in which each actor can compensate for the weaknesses of others.

International governing agencies have the financial capital and technological resources to fund training programs and gather information about environmental changes. Often communities in the south lack these resources. City governments in Africa, Asia and Latin America often have one hundredth or (in their most extreme) one thousandth of the revenue per capita available to most city or municipal governments in Europe (Hardoy, Mitlin and Satterthwaite, 1992, p. 162). In spite of this, these cities have similar responsibilities that are possibly even greater due to poverty and environmental degradation. Therefore, outside assistance is necessary to provide additional financial capital and technological resources.

The power and global perspective of international governing agencies also facilitates their role as legislators in creating legally-binding instruments for environmental protection and resource management. Although legal agreements are fundamental in changing inappropriate practices of nations and industries, they may be difficult to enforce due to lack of human resources in some regions.

The main criticism of top-down strategies is that they often do not share the same perspective of sustainable development as local communities. International institutions' liberal approaches may actually preclude sustainable development by advocating wider-scale market economies that will encourage more exploitation of resources. To counter this force and safeguard the resources of local communities, bottom-up grass roots movements need to be integrated into the coastal management process.

It is from the bottom-up aspect of ICM that NGOs and social movements emerge as counter weights to international institutions and corporations. Hillary Clinton (1995) aptly describes the role of NGOs as counter weights: "It is the NGOs who have pressured governments and led governments down the path to economic, social, and political progress, often in the face of overwhelming hostility" (p. 674). NGOs have an important role in planning policies, networking, implementation, and monitoring of programs. Perhaps, though, the most important role NGOs have is the raising of consciousness about environmental issues and educating citizens about more appropriate behaviors for sustainable development. Outside the formal education system, NGOs can promote local, experiential non-formal education. Many proponents of environmental education affirm the workability of eco-centered educational programs in non-formal education (van Riezen, 1996; Taylor, Hadsell, Lorentzen and Scarce, 1993). Eileen B. Claussen (1997), U.S. Assistant Secretary of State for Oceans and International Environmental and Scientific Affairs, describes the need for this education:

> Locally, education is absolutely critical. In our increasingly urbanized societies, it is all too easy to forget how our daily activities affect the marine environment. The way we change the oil in our cars, the methods by which we apply pesticides in the garden, the choices we make about using toxic

substances in our homes and the way we dispose of those substances—all of these actions end up affecting marine ecosystems. Changing our habits in response to environmental concerns requires massive education at the local level, yet the result of such an effort is sure to be a substantial reduction in the flow of pollutants into the marine environment (Effective implementation section, para. 2).

In his book, *Earth in the Balance*, former vice president Al Gore (1992) also advocates education to develop more eco-centric views in his proposal of a "Global Marshall Plan" for the environment. One of his strategic goals is "the establishment of a cooperative plan for educating the world's citizens about our global environment" (p. 306). He reasons that such a plan could "foster new patterns of thinking about the relationship of civilization to the global environment" (p. 307). These new patterns of thinking can give rise to change just as social movements incited new ways of thinking about communism in Eastern Europe led to many political changes. Unfortunately, the extent of Gore's advice for educational reform is merely to use technology for retrieving and passing on information about environmental concerns. Although this can foster an awareness that is vital in changing attitudes; more is needed to create a workable and effective environmental education curriculum in schools around the world.

Consideration of future generations is a key element in environmental education as C.A. Bowers (1995) notes in his discussion of trans-generational communication in the educational process. Bower expresses the need to shift away from student-centered learning and toward a process "of encoding, storing, and renewing a cultural group's ways of understanding and valuing the primary life sustaining relationships between humans and the rest of the biome" (p. 135). This is an eco-centered approach that emphasizes tradition and culture in a way that will require the elder generations to act as "carriers of essential knowledge and values". The environmental education process that Bower describes echoes with words like "stewardship", "nurturing" and "emancipatory education".

Education can also take place in the form of social movements that aim at assembling critical masses to gain media attention and further raise public awareness of their campaigns. Alberto Melucci (1996) explains this process:

> Movements touch upon the most sensitive mechanisms of society and they come to coincide with the frontiers of change; both as expressions of the forces of resistance ("reactionary" and conservative forms of opposing change) and in the form of innovation (creative social initiatives) the issues raised by social movements are always crucial for a given society as a whole. They are in effect indicators of what is taking place in a given society, and today they reveal in many ways the core processes of national societies and our global system alike (pp. 146-147).

A New Environmental Paradigm

Education and social movements can be driving forces in developing a new social paradigm that views the environment as an important resource that needs to be protected to ensure a better quality of life for future generations. Environmental education is emancipatory in that it leads to the reproduction of environmental values. The passing on of environmental values from one generation to another begins the process of structuring a new social paradigm. Environmentally based non-formal education can change the way people think about their relationship with nature. Lester Milbrath (1989) aptly argues for the need to promote new social paradigms that focus on sustainability and reconsider the way society dominates the environment. Some of Milbrath's ideas are radical in that they require a massive restructuring of political institutions and society. Milbrath also advocates a shift toward placing a higher valuation on nature, carefully planning action to avoid risks and limiting growth.

Through education and consciousness raising, non-formal education can help citizens realize the dependence humans have on the environment. This could create a more holistic perspective that tightens the relationship between humans and nature. The ultimate goal here, however, is to encourage behavior that favors environmental protection over economic growth. Economic growth is not necessarily harmful; this simply means that environmental protection should be a priority. To maintain a balance, careful planning is needed. Planning should consider all short-term and long-term risks. Education is an important element in the planning process because knowledge allows communities to make informed decisions about their lives. A crucial element of informed planning is the ability to realize the limits of growth. Thus, one major goal of integrated coastal management programs is to determine what types of growth could lead to the sustainable development.

References

Archer, D. (1994). "The Changing Roles of Non-Governmental Organizations in the Field of Education" (in the context of changing relationships with the state). *International Journal of Educational Development,* 14(3): 223-232.

Bower, B.T., Ehler, C.N. and Basta, D.J. (1994). *A Framework for Planning for Integrated Coastal Zone Management.* Silver Spring, MD: NOAA/NOS Office of Ocean Resources Conservation and Assessment.

Bowers, C.A. (1995). *Educating for an Ecologically Sustainable Culture.* Albany, NY: State University of New York Press.

Claussen, E.B. (1997, January). "Critical Coastlines", *Our Planet,* 8.5. Retrieved from http://www. ourplanet. com/imgversn/85/claussen. html

Clinton, H. (1995). "NGO Forum: Advancing the Progress of Women", *US Department of Dispatch,* 6(9): 672-674.

Cruz, W. and Munasinghe, M. (1996). "Environmental Implications of Economic Policies", *Environment,* 38(6): 6-11.

Empowering Women: More Education, Better Health Care, Less Poverty (1995). *UN Chronicle,* 32(6): 46-47.

Fernandes, R.C. (1985). *Non-Governmental Organizations (NGOs), A New Institutional Reality in Latin America.* Rome: Food and Agriculture Organization of the United Nations.

Gore, A. (1992). *Earth in the Balance.* New York: Plume.

Hallak, J. (1990). *Investing in the Future: Setting Educational Priorities in the Developing World.* Paris: UNESCO-IIEP.

Hardoy, J.E., Mitlin, D. and Satterthwaite, D. (1992). *Environmental Problems in Third World Cities.* London: Earthscan Publications.

Imber, M.F. (1994). “The United Nations’ Role in Sustainable Development”, in C. Thomas (ed.), *Rio: Unraveling the Consequences*. Portland, OR: Frank Cass and Co.

Independent World Commission on the Oceans (1998). *The Ocean, Our Future*. Cambridge, England: Cambridge University Press.

McCormick, J. (1993). “International Non-Governmental Organizations: Prospects for a Global Environmental Movement”, in S. Kamieniecki (ed.), *Environmental Politics in the International Arena: Movements, Parties, Organizations, and Policy*. Albany, NY: State University of New York Press.

Melucci, A. (1996). *The Playing Self: Person and Meaning in the Planetary Society*. Cambridge, England: Cambridge University Press.

Milbrath, L.W. (1989). *Envisioning a Sustainable Society: Learning Our Way Out*. Albany, NY: State University of New York Press.

National Oceanic and Atmospheric Administration (2000). “Integrated Coastal Management: An International Priority”, Retrieved from http://icm. noaa. gov/story/icm_intro. html

Schubert, L. (1993). “Environmental politics in Asia”, in S. Kamieniecki (ed.), *Environmental Politics in the International Arena*. Albany, NY: State University of New York Press.

Taylor, B., Hadsell, H., Lorentzen, L. and Scarce, R. (1993). “Grass-Roots Resistance: The Emergence of Popular Environmental Movements in Less Affluent Countries”, in S. Kamieniecki (ed.), *Environmental Politics in the International Arena*. Albany, NY: State University of New York Press.

Thai-Eng, C. (1998). “Lessons Learned from Practicing Integrated Coastal Management in Southeast Asia”, *Ambio*, 27(8): 599-610.

Toh, S. and Floresca-Cawagas, V. (1997). “Towards a People-Centered Education: Possibilities and Struggles in the Philippines”, *International Review of Education*, 43(5/6): 527-545.

United Nations Conference on Environment and Development (1992). *Agenda 21: Chapter 17 - Protection of the Oceans, All Kinds of Seas, Including Enclosed and Semi-Enclosed Seas and Coastal Areas and the Protection Rational Use and Development of their Living Resources*. Retrieved from http://www.unep.org/Documents/?DocumentID=52&ArticleID=65

van Ginkel, H. (1998). *Preparing for a Sustainable Future: Higher Education and Sustainable Human Development*. Retrieved from http://www.unesco.org/iau/fre/tfsd_sustain1.html

van Riezen, K. (1996). “Non-Formal Education and Community Development: Improving the Quality”, *Convergence*, 29(1): 82-95.

Viswanath V. (1991). *NGOs and Women's Development in Rural South India: A Comparative Analysis*. Boulder, CO: Westview Press.

Voabil, C., Engdahl, S. and Banze, J. (1998). "SEACAM: Coastal Zone Management in Eastern Africa has Taken a New Approach", *Ambio*, 27(8): 776-778.

Warren, K. (1996). "Ecological Feminist Philosophies: An Overview of the Issues", in K. Warren (ed.), *Ecological Feminist Philosophies*. Bloomington, IN: Indiana University Press.

Williams, M. (1996). "International Political Economy and Global Environmental Change," in J. Vogler and M.F. Imber (eds.), *The Environment and International Relations*. New York: Routledge.

World Bank (1992). *World Development Report 1992*, New York: Oxford University Press.

World Commission on Environment and Development (1987). *Our Common Future*. Oxford: Oxford University Press.

Environment-Economy Trade-offs and Forest Environmentalism

8

Carl M. Hand, Ginger Macheski

Public opinion polls have consistently demonstrated the public's willingness to "trade-off" economic growth for environmental protection. The absence of a middle ground position in these polls raises questions as to whether or not respondents view economic growth and environmental protection as necessarily inimical. Furthermore, little research has examined whether these trade-off preferences actually relate to how individuals view environmental problems at the local level. This research examines the trade-off problem in light of a current resource conflict, forest management, using a statewide probability telephone survey. Most respondents prefer to see that "both the environment and the economy" are important in the case of forestry management, but favor weighting the balance in favor of environmental protection. Furthermore, general trade-off preferences correlate with specific value choices (relating to private property), definitions of the current forestry situation for the state, and attitudes toward development of forest resources. At least in the case of forest management, how individuals conceptualize the economy/environment trade-off is central to how they think about forest problems more generally.

Introduction

A central question in environmental debate today is how environmental protection and economic growth interrelate. As Schnaiberg (1980) notes, economic growth is dependent on "biospheric withdrawals" and a "treadmill of production" that does not easily accommodate environmental protection. The complexity of the environmental-economic relationship, however, doesn't lend itself to simple generalizations. There is a great deal of evidence that environmental protection and economic growth aren't necessarily antithetical (Schug and Western, 1997) and even that economic growth is necessary for pollution abatement (Byrne, 1997). Some analysts have begun to make headway in decoupling these issues. For example, Lovins, Lovins and Hawken (1999) argue that reducing demand for energy (e.g. through energy efficient motors) may be essential for continued economic growth. As Hertsgaard (1998, p. 300) notes, environmental innovations that save money (and resources) will become an important way to generate new jobs. In short, it is not unreasonable to assert that economic systems can be made more both more environmentally benign *and* more profitable. Increasing resource efficiency in the future, however, may be entirely offset by the growth and expansion of human population with its attendant demand for resources.

Pollsters have attempted to tap public loyalties in this debate through the so-called "trade-off" question. The typical form of this question has been to present respondents with dichotomous alternatives between favoring environmental protection or favoring economic growth or jobs. Longitudinal trends in the United States (as documented in publications such as *Cambridge Reports Trends & Forecasts* by Cambridge Reports, Inc.) indicate growing support for the environmental choice implying, for many, decreasing commitment to traditional economic and materialistic values (Inglehart, 1995). The public's apparent willingness to promote environmental quality over economic growth is often viewed as evidence of deepening environmentalism (Dunlap, 1991).

While trade-off questions provide some indication of changing public values about the environment, their abstract as well as dichotomous format raises many issues regarding how the public views the relationship between economic growth and environmental protection. Are they seen as necessarily inimical, requiring that we trade off one against the other? Conversely, might the public see a healthy economy and a healthy environment as desirable end-states of social policy, both of which can be equally maximized? Furthermore, do trade-off questions implicate how individuals think about environmental problems? Do individuals think of the relationship in the same way when dealing with local conflicts between the environment and the economy? This paper examines these issues in light of a current environmental conflict, forest management.

Overview

Poll data in the United States indicate growing support for the environmental choice in environment-economy trade-offs (Dunlap, 1991; Kempton, Boster and Hartley, 1995). Perhaps the best long-term evidence for this can be seen in the *Cambridge Reports*, which asked a form of the trade-off question between 1976 and 1996.[1] The question has two choices and includes a "don't know" response. The trend in these data indicates an increasing percentage of respondents who are willing to sacrifice economic growth for environmental protection. For example, between 1976 and 1993 the percentage willing to "sacrifice economic growth in order to preserve and protect the environment" increased from 39% to 64%. The percentage of those willing to "sacrifice environmental quality for economic growth" on the whole does not vary widely, ranging between 15% and 32%. Between 1977 and 1996 the percentage indicating, "don't know" was almost halved, going from 41% to 24%.

Like *Cambridge Reports*, Gallup has also asked a form of the trade-off question using two response categories: "Protection of the environment should be given priority, even at the risk of curbing economic growth; or economic growth should be given priority, even if the environment suffers to some extent." The question was asked in 1984 and intermittently between 1990 and 2001. The data is consistent with *Cambridge Reports*. Between 65% and 71% of the respondents favor environmental protection over the decade of the 1990s, similar to the 61% who favored environmental protection in 1984. Over the entire period the percentage indicating no opinion was cut in half, going from 11% in 1984 to 5% in 1999.

We get a slightly different picture of the trade-off, however, when the question switches to a real policy debate. Ladd and Bowman report that in a survey by the Wirthlin Group in 1994, three quarters of respondents selected the response "it does not necessarily have to be a choice between the two". Thus, most Americans probably see very little antagonism between economic growth and environmental protection in the abstract. For example, Gallup found in 1991 that 78% of respondents supported taxing corporations who produced hazardous waste more, even if this resulted in higher prices on consumer goods. Yankelovich found a threefold increase between 1984 and 1990 in the public's willingness to spend more on consumer goods if this assured that business would not harm the environment (Kempton et al., 1995, p. 5). How much more individuals would be willing to pay is another test of how much the public is willing to trade off economic prosperity for environmental protection. For example, Gallup found in 2001 that 75% of the public is willing to pay $ 100 more in prices for consumer goods each year if this would help reduce pollution. When the dollar value is increased to $ 500, 61% of respondents are willing to agree. Thus while the public is willing to make economic sacrifices, the extent of these sacrifices may be quite limited.

One additional feature of the trade-off question can tell us something about how the public views the relationship between the environment and the economy; that is, the "don't know" response. Ladd and Bowman (1995) note that the *Cambridge Reports* data have high percentages of respondents who don't know—as high as 42% in 1983. This would suggest that the trade-off idea is problematic for the public. As Ladd and Bowman note, the inability to chose between the two alternatives "could reflect ignorance, or ... indicate that the choices presented do not capture public thinking on the problem" (p. 24). Ladd and Bowman report that in a survey by the Wirthlin Group in 1994, fully 74% of respondents picked a third choice—"it does not necessarily have to be a choice between the two", and only 15% said economic growth need be sacrificed. Thus the public may not see environmental protection as inimical to economic growth—that it is a false choice. Given the general lack of attention to environmental issues, most Americans probably see very little antagonism between economic growth and environmental protection in the abstract. As Ladd and Bowman note: "Because Americans firmly believe that we can have both a clean environment and economic growth, questions that ask Americans to choose between one and the other are highly misleading" (p. 24).[2] The view thus emerging from the research conducted thus far is that public expects to achieve both economic growth *and* environmental protection. Brechin and Kempton (1994) state this succinctly when they argue that "rather than being a postindustrial luxury, a healthy environment is now widely viewed as essential for a sound economic base" (p. 265).

The potential conflict between economic growth and environmental protection becomes more apparent to respondents when they are asked to view the trade-off in light of their own local resource issues. This conflict is evident in the poll data already discussed regarding how trade-offs might affect the respondent's standard of living or local unemployment. One view would suggest that attitudes toward environmental protection and economic growth both tap into larger quality of life concerns (McBeth, 1995). O'Connor, Bord and Pflugh (1994) support this view in their study of a New Jersey community's opposition to coastal economic development. They critique the notion that environmentalism implies hostility toward economic growth. Environmentalism at the local level may be more of an expression of concern for *how* development is occurring—declining green space, misuse of resources, and population growth—rather than opposition to economic growth *per se*.

This research points out the necessity for understanding the question of trade-offs in light of particular development or resource issues. Most national studies have been unable to do this given the unique circumstances of these issues. This study examines the trade-off problem in light of questions relating to forest management. We believe forestry management presents a good case study for how the public thinks about trade-off questions. Forestry management issues have come under increased public scrutiny, often eliciting emotional debate (Fortmann and Kusel, 1990). This has been particularly evident in the old growth forest debates in the Pacific Northwest. As Perry and Pope (1995) note, forestry debates tend toward polarization. At root, these conflicts emerge from the fundamental value and ethical divide that views forest resources as commodities to be harvested for human uses or as natural resources to be preserved for their ecological, esthetic and recreational value. This tendency toward either economic utilitarianism or environmentalism in forestry debates makes them ideal for understanding economy-environment trade-offs more generally. This study examines trade-off preferences within the area of forestry management using a statewide survey of Georgia residents. In particular, we will examine whether or not trade-off preferences relate to more specific attitudes toward forest protection and forest development.

Methods

In order to examine these issues, a statewide random sample of Georgia residents 18 years and older conducted by phone between April and May in 1996 will be used. The sample was obtained from Survey Sampling, Inc. in order to assure representativeness. Telephone interviews were supervised and coordinated at Valdosta State University. The total number of completed surveys is 844, representing 71% of the original sample size. The sampling error is approximately 3% at the 95% confidence level. The survey was specifically designed to look at issues and attitudes pertaining to forestry management and resources. A number of other questions are included in the study, including environmental trade-offs, attitudes toward specific components of forestry management, and selected demographic variables.

The review of literature suggests that environmental trade-offs present respondents with what they may perceive to be a false dichotomy, forcing fence-sitting respondents into one or the other alternative or the "don't know" category. Providing respondents more of a continuum of choices will alter this outcome, providing us a truer picture of how individuals view such trade-offs. In addition, trade-off questions ask respondents to think about the relationship between the environment and the economy in the abstract; are they equally supportive of environmental protection when considering specific resource issues, such as forestry management? Furthermore, do the different trade-off preferences correlate with more specific opinions of forestry management?

For this study, the trade-off question asks individual to think about the relationship between environmental protection and economic growth in light of forest resources. The environmental trade-off question is asked in the following way: "Many forest management issues involve difficult trade-offs between environmental and economic considerations. Which of the following statement best describes your view?

1. The highest priority should be given to protecting the environment, even if it hurts the economy.
2. Both the environment and the economy are important but the environment should come first.
3. Both the environment and the economy are important but the economy should come first.
4. The highest priority should be given to economic considerations such as jobs even if it hurts the environment.

Other questions are included in the survey for examining a respondent's definition of the forestry situation in Georgia, including overall concern for forests, the degree of tree cutting, government regulation of forestry resources, private property rights, and specific forest development activities.

Results

The trade-off question provides for extremes in the environment-economy trade-off as well as the two middle categories that affirm the belief that "Both the environment and the economy are important". The vast majority of respondents have picked the middle ground, 72.1% in the second group (tilting the balance toward the environment), and 13.3% in the third group (weighing in on the economic side). A total of

85.4% saw both the environment and the economy as important; a percentage not remarkably different from the Wirthlin Survey in 1994 alluded to earlier. Only 5.3% favor jobs to the detriment of the environment. A comparable number of respondents, 8.5%, believe that protecting the environment is of paramount importance, even at the risk of hurting the economy. Unlike previous poll data, virtually none of the respondents (less than 1%) picked the "don't know" category. This is undoubtedly due to the fact that respondents were given a range of responses rather than an either/or choice.

Overall, these data attest to the enormous popularity of the notion of balancing environment and economic interests, but weighting those interests more heavily on the environmental side. As we noted earlier, a dichotomous trade-off question fails to capture how most individuals think about the relationship between the environment and the economy. The evidence here would suggest that respondents reject a polarized definition of the problem. Respondents obviously value both economic growth and environmental protection within the area of forestry management; if a trade-off has to be made they would prefer that it be for environmental protection.

Earlier, we raised the question of whether or not respondents' trade-off preferences implicate more specific attitudes toward local resource or development issues. If the apparent preference for environmental protection over economic growth reflects merely the current socially desirable response, we would not expect it to correlate strongly with these more specific attitudes. On the other hand, trade-off preferences might emerge from how respondents view environment-economy relationships at the local level and are thus more likely to be linked to specific attitudes and beliefs. We think this is relevant to understanding how the public engages in debates regarding real environmental or economic trade-offs.

Table 1 examines this problem by looking at a central issue in forestry management—the role of private property rights. This table cross tabulates the trade-off (Trade off) question with four questions on government regulation of private land. Overall, the data suggest that those more favorable to the environment side of the trade-off are also more supportive of efforts to regulate private land in order to protect the environment. The biggest percentage difference between groups is in the last question (Private), and is statistically significant, as are Property and Limit. Interestingly, the majority of respondents in the most pro-economy groups (last two groups) favor government regulation of private land on behalf of the environment. Furthermore, there is little difference between the two pro-economy groups in terms of their support of private property rights. Thus, the evidence would suggest that the majority of respondents in the pro-economy groups should be viewed as supportive of environmental protection even when this conflicts with private property rights, but to a lesser degree than the pro-environment groups.

Table 1: Trade-off by Private Property Rights

Question	Protect Env. Hurt Econ. % Agree	Both Important Favor Env. % Agree	Both Important Favor Econ. % Agree	Protect Econ. Hurt Env. % Agree	X^2
Compensate[1]	75.9	75.6	80.9	76.8	8.8 (p=ns)
Property[2]	92.3	91.9	77.7	78.0	37.7 (p<.001)
Limit[3]	87.9	89.0	65.6	70.8	58.4 (p<.001)
Private[4]	37.3	27.2	53.2	62.5	51.9 (p<.001)

N=743

1. *Compensate*: "If a forest landowner is prevented from cutting trees on his land because of government regulation, the landowner should be paid for the economic loss."
2. *Property*: "Private property rights are important but only if they don't hurt the environment."
3. *Limit*: "Private property rights should be limited if necessary to protect the environment."
4. *Private*: "Private forest owners have the right to do as they please with their forests regardless of what it does to the environment."

Table 2 reports the cross tabulation of trade-off with specific definitions of the forestry situation in Georgia. First, 49.1% of respondents picking the most pro-environment trade-off indicate that they have concerns about forests in Georgia compared to only 19.4% of those in the most pro-economy group, a significant difference ($X^2 = 14.8$, $p = .003$). Also, a significant difference exists between respondents with respect to perceptions of tree cutting in Georgia; the majority of pro-environment respondents would like to see less tree cutting, while only 23.5% and 30.6% of the two pro-economy groups respectively feel the same way. Finally, no difference exists between trade-off groups regarding perceptions of forestry owners. These data suggest that pro-environment respondents view the forestry situation in more problematic terms than pro-economy respondents do.

Table 2: Trade off by Definitions of Forestry Problems

Question	Protect Env. Hurt Econ.	Both Important Hurt Env.	Both Important Favor Env.	Protect Econ. Hurt Econ.	X^2
Concerns about Forests[1] %Yes	49.1	47.6	33.7	19.4	14.8 (p.<.003)
Tree Cutting[2]					
More Tree Cutting	3.1	2.4	4.9	2.8	28.6 (p<.001)
Less Tree Cutting	53.1	49.4	23.5	30.6	
About Right	43.8	48.3	71.6	66.7	
Opinion of Timber Owners[3] %Favorable	43.3	46.9	59.6	63.9	15.9 (p=.ns)

1. "Do you have any concerns about the way forests in Georgia are being treated?"
2. "In general, do you think that there should be more tree cutting, less tree cutting, or is it about right?"
3. "Do you have a favorable or unfavorable opinion of the timber land owners in Georgia?"

Table 3 cross-tabulates each of the trade-off groups by the percentage favoring (favor and strongly favor) specific development strategies of forestry resources. Overall, the first two groups are less favorable to forestry-based development activities in Georgia than the last two groups. These differences are statistically significant in four of the six types of activity. For developing tourism, statistical significance exists between those respondents choosing the most pro-environmental response ("hurt economy") and everyone else. Still, the lowest percentage favoring a particular activity is 59% for the first trade-off group. Thus while pro-environment groups are less supportive of specific forest developments, the majority of respondents favor these activities.

Table 3: Trade-off by Percent Favoring Specific Forestry Development Activities[1]

Question	Protect Env. Hurt Econ. % Favouring	Both Important Favor Env. % Favouring	Both Important Favor Econ. % Favouring	Protect Econ. Hurt Env. % Favouring	X^2
Constructing Lumber Mills	60.7	73.3	84.0	88.9	20.5 (p<.03)
Exporting Lumber	59.4	60.6	76.3	65.7	14.9 (p=.ns)
Developing Tourism	66.1	87.3	88.0	86.5	32.1 (p.<.001)
Exporting Wood Chips/Logs	67.7	65.0	80.0	76.9	16.7 (p.=ns)
Constructing pulp/paper mills	69.3	69.9	83.1	90.2	31.9 (p.<.001)
Constructing chipmills	78.3	77.9	88.8	89.8	21.9 (p.<.05)

1. "Future economic development in Georgia could potentially include growth of the forest industry. I want to read some types of economic development activities to see which you would prefer, assuming that they are carried out under good forestry practices."

Overall, these results point to several findings. First, the majority of respondents prefer a balanced approach to the environment-economy trade-off. Very few respondents wish to pursue environmental protection or economic growth to the detriment of the other. Second, significant differences were found between trade-off groups with respect to their attitudes toward forest protection. Pro-environment respondents (first two groups) tend to be less supportive of private property rights when these conflict with forest protection and more supportive of government regulation of private property than pro-economy groups (second two groups). Difference between the trade-off groups was one of degree, however, given that the majority of respondents favor these things. To put it differently, pro-economy groups are not hostile to efforts to ameliorate environmental problems. We also found significant differences between groups in terms of their definition of the forestry situation in Georgia. Pro-environment groups have a more problematic view of Georgia forests, that is, they were more willing to voice concern about forests and view logging as excessive and to support policies to protect forests. These attitudes result in significant differences regarding support for some forestry development activities in the state.

Conclusion

Environmental protection and economic growth are often portrayed in public debates in zero-sum terms, particularly at the local level (Clarke and Moss, 1990). As Hertsgaard (1998) has argued: "The assumption that environmental protection must cost jobs and lower profits ... has been repeated for so many years by so many people ... that it has come to be regarded as a fact" (p. 298). While policy debates have tended toward polarization—a "pro-growth" or "pro-environment" world view—public opinion is more complex. On the one hand, solid majorities support trading off economic growth for environmental protection and appear ready to accept the costs of doing so via increased taxes or prices for consumer goods. But the data also suggest that the public's willingness to do so drops off fairly quickly when dollar values rise above token amounts. This implies a tacit realization that reducing economic growth with its attendant job loss or redistributing economic growth through higher taxes and prices can impact individual quality of life in negative ways. Thus, while environmental protection AND economic growth may be viewed by many as complementary public goods in the abstract, the balance of these goods will vary significantly at the individual level.

Data reported here suggest that most respondents are neither exclusively pro-environment nor pro-growth. Rather, they perceive the need to balance these concerns when dealing with forest resources. The majority of respondents favor environmental protection while recognizing the continuing need for economic growth. While this may seem like an easy out, respondents are willing to sacrifice private property rights, reduce the amount of logging, and restrict forestry developments in order to accomplish this. This suggests not an antithesis to economic interests, but a realization that environmental protection is less likely to be factored into the equation when short-term economic interests are at stake. Furthermore, many view trading-off individual and economic interests in the short-run for environmental protection as the means to sustainable economic growth. This is precisely the view emerging in many environmental circles (e.g. Lovins et al., 1999). The conservation and wise-use of natural resources has always been at least a part of American culture, particularly in rural populations (McBeth, 1995). Such values complement rather than conflict with a desire for economic growth. It follows that policy leaders should pursue a strategy in which environmental protection and increasing resource efficiency are viewed as the *means* to enhancing future economic growth, rather than a strategy of zero-sum.

Endnotes

1. *Cambridge Reports* Trend Data (Earl Taylor, personal communication, January 15, 1999):

	1976	1977	1978	1979	1981	1982	1983	1984	1985	1986	1987
Sacri. Env. (%)	21	26	23	32	26	31	16	27	23	19	23
Sacri. Econ (%)	38	39	37	37	41	41	42	42	53	58	57
DK (%)	41	35	39	32	33	28	42	31	24	23	20

	1988	1989	1990	1991	1992	1993	1994
Sacri. Env. (%)	19	21	15	18	19	24	23
Sacri. Econ (%)	52	52	54	63	56	64	53
DK (%)	29	27	22	19	25	22	24

2. Ladd and Bowman add that when we must take into account that while many people express a growing concern for environmental problem, these issues are not highly salient to them personally. Environmental issues have more or less dropped off the "most important problem" list (which are typically elicited with an open-ended question). This lack of salience is ironic given other poll data showing that Americans strongly identify with the environmentalist label (as many as a quarter of the public would identify themselves as such according to a 1987 Times Mirror study).

References

Brechin, Steven R. and Kempton, Willett (1994). "Global Environmentalism: A Challenge to the Postmaterialism Thesis", *Social Science Quarterly,* 75(2): 245-269.

Byrne, Margaret M. (1997). "Is growth a Dirty Word? Pollution, Abatement and Endogenous Growth", *Journal of Development Economics,* 54, pp. 261-284.

Clarke, Susan E. and Moss, Anne K. . (1990). "Economic Growth, Environmental Quality, and Social Services: Mapping the Potential for Local Positive-Sum Strategies", *Journal of Urban Affairs,* 12(1): 17-34.

Dunlap, Riley E. (1991). "Trends in Public Opinion Toward Environmental Issues: 1965-1990", *Society and Natural Resources,* 4, pp. 285-312.

Fortmann, Louise and Kusel, Jonathan (1990). "New Voices, Old Beliefs: Forest Environmentalism Among New and Long-Standing Rural Residents", *Rural Sociology,* 55(2): pp. 214-232.

Gallup, George, Jr. (1984). *The Gallup Poll: Public Opinion 1984.* Wilmington, DE: Scholarly Resources.

Gallup, George, Jr. (1990). *The Gallup Poll: Public Opinion 1990.* Wilmington, DE: Scholarly Resources.

Gallup, George, Jr. (1991). *The Gallup Poll: Public Opinion 1991.* Wilmington, DE: Scholarly Resources.

Gallup, George, Jr. (1995). *The Gallup Poll: Public Opinion 1995*. Wilmington, DE: Scholarly Resources.

Gallup, George, Jr. (1998). *The Gallup Poll: Public Opinion 1998*. Wilmington, DE: Scholarly Resources.

Gallup, George, Jr. (1999). *The Gallup Poll: Public Opinion 1999*. Wilmington, DE: Scholarly Resources.

Gallup, George, Jr. (2001). *The Gallup Poll: Public Opinion 2001*. Wilmington, DE: Scholarly Resources.

Hertsgaard, Mark (1998). *Earth Odyssey: Around the World in Search of Our Environmental Future*. New York: Broadway Books.

Inglehart, Ronald (1995). "Public Support for Environmental Protection: Objective Problems and Subjective Values in 43 Societies", *PS: Political Science and Politics,* 28(1): 57-72.

Kempton, Willett, Boster, James S. and Hartley, Jennifer A. (1995). *Environmental Values in American Culture*. Cambridge, MA: The MIT Press.

Ladd, Everett Carl and Bowman, Karlyn H. (1995). *Attitudes Toward the Environment* (AEI special studies in public opinion). Washington, DC: American Enterprise Institute for Public Policy Research.

Lovins, Amory B., Lovins, L. Hunter and Hawken, Paul (1999). *Natural Capitalism: The Coming Efficiency Revolution.* New York: Little Brown.

McBeth, Mark K. (1995). "Rural Environmental and Economic Development Attitudes: An Empirical Analysis", *Economic Development Quarterly,* 9(1): 39-49.

O'Connor, Robert E., Bord, Richard J. and Pflugh, Kerry Kirk (1994). "Two Faces of Coastal Environmentalism: Environmental Protection and Development on Cape May", *Coastal Management,* 22: pp. 183-194.

Perry, Gregory M. and Pope, C. Arden III (1995). "Environmental Polarization and the Use of Old-Growth Forests in the Pacific Northwest," *Journal of Environmental Management,* 44: pp. 385-397.

Schnaiberg, Allan (1980). *The Environment: From Surplus to Scarcity.* New York: Oxford University Press.

Schug, Mark C. and Western, Richard D. (1997). "An economic Perspective on Protecting the Environment", *Social Education,* 6(6): pp. 329-330.

Production System Planning for Natural Resource Conservation in a Micro-Watershed

9

Nallathiga Ramakrishna

Natural resources management is crucial for sustainable development of a region. It involves the knowledge of natural resources, planning for the sustainable use, and implementation at field level. A watershed is an area within a region with natural boundaries that are most appropriate for natural resources management. This paper emphasizes the need for a watershed approach, provides a brief description of that approach taken by India, and presents an exercise in planning and aiding in decision-making for a case study watershed. Production Systems Planning (PSP) is an approach presented here that can be taken for achieving natural resource conservation for the case study watershed. The proposed conservation measures are quite useful in the harmonious development of the region for food production.

Introduction

Man's interaction with nature is rapidly becoming more complex due to a multitude of activities that directly or indirectly cause a disturbance in the natural system. The deeper interactions between human activities and natural ecosystems call for an interdisciplinary approach to natural resources management, while the inputs from multiple disciplines need to be effectively utilized in achieving it. This prudent natural resources management will enable sustainable development of a region without losing the resource base. Sustainable development essentially aims to reconcile conflicting objectives of economic development and improvement in human welfare, and ecological sustenance and functioning of ecosystems. The term *sustainable development* is defined by the Brundtland Commission as "the development that constitutes meeting needs of the current generation without compromising the ability of meeting needs of future generations" (Brundtland, 1989, p. 43). Sustainable development is now globally accepted as a supreme long-term goal of humanity (World Commission on Environment and Development, 1987). Sustainable utilization of natural resources is essential for sustainable development.

It follows that the sustainability of renewable natural resources in general and common property resources in particular assumes importance in both developed as well as developing countries because of the finite capacity of the resource base and the increasing demand for its exploitation. The technical definition of the term *sustainability* is given as, "the ability of a natural resource system to produce a socially optimum level of output that is necessary to meet the needs and aspirations of the people dependent on the system perpetually without any detrimental effects on the resource system itself and the physical environment, and with no imposition of significantly greater risks on future generations" (Colchester as quoted by Singh, 1995, p. 1). This might be comprehensive and deep rooted, but in other words, sustainability implies not only conserving natural resources but also maintaining ecological functions and the supply of natural resource products, which are essential to the livelihoods of local people. Sustainability in this sense is a dynamic concept that reflects changing levels of output corresponding with changing human needs and production technologies over time.

Natural Resources Management

Natural resources discussed here are broadly covered under land, water and biomass (or vegetation). These three resources are crucial for production under various systems namely, agriculture, horticulture, silviculture, pisciculture and animal husbandry. Management of these three major resources is crucial for making production in these systems sustainable and enhanced. Sustainability of these systems can be threatened due to disruption in the linkages between these resources. This may assume various forms, for example, increased soil erosion resulting in nutrient loss rendering it unfit for use in the case of land; soil water deficiency or excessiveness affecting its productivity and degradation in the case of water resources; decline in vegetation density and diversity leading to reduction in soil and water conserving properties in the case of biomass or vegetation resources, and so forth. A more detailed discussion of these follows:

Soil

Soil productivity can be defined as those properties of soil (physico-chemical and biological) that influence crop production. The increased yields from better-managed soils are due to increased inputs and improved practices rather than with improvements in the basic fabric of soil (soil productivity). In recent years the sediment derived from soil erosion has been the major non-point source of pollution in surface water bodies while loss of *in situ* topsoil has caused reduction in productivity. Erosion reduces long-term production potential and seldom improves the immediate capacity of eroded soil to sustain plant growth or produced crops (Pierce, Larson, Dowdy and Graham, 1983). Results of recent studies show that soil's physical and biological properties seem to be the predominant constraints to maximizing plant production on eroded soil, compared with chemical and fertilizer constraints. For example, Rosenberry, Knutson and Harmon (1980) suggested that yields generally decline as soils shift from one erosion phase to another, even with increased fertilizer. This is attributed to surface soil physical and biological properties. However, in addition to irrigation, other water management techniques, such as surface mulching, can also be applied for amelioration of eroded soil.

Water

Crop productivity is also affected by moisture availability in soil. In the post-green revolution, particularly in the 1980s, instability in crop productivity increased on account of the rise in sensitivity of output to variations in rainfall in India (Rama Mohan Rao et al., 1999). This increasing vulnerability of agricultural output to variations in rainfall, particularly during droughts when the soil moisture is scarce, is attributed to inadequate expansion of irrigation by these same authors. It is minor irrigation, which is not given priority that can be part of the strategy under watershed development. Similarly, decline in water quality has affected crop productivity in saline and alkaline lands that were created by excessive irrigation or polluted water in northern parts of India.

Vegetation

At the same time, vegetation adds much to resource endowment and has crucial linkages with soil and water. Good vegetation cover functions as a soil and water-conserving agent, whereas, lack of vegetation will make the soil vulnerable to erosion and allow water to flush off the sediments. Biological diversity of vegetation is crucial in the survival of the vegetation itself and the sustenance of the ecosystem in that region. In fact, planning based on land use can effectively conserve soil as well as water. This is little elaborated under Production Systems Planning (PSP).

A distinction, however, needs to be made between the goals of attaining sustainability and of increasing productivity. While higher productivity may be required to achieve the sustainability goal, the requisite increase in productivity must be achieved in a manner that will not jeopardize the ability of a natural resource system to meet future needs. In other words, it is possible to achieve increases in productivity through unsustainable short-term approaches (York as cited in Singh, 1995).

Watershed Management

The term *watershed* denotes the area defined by natural boundaries characterized by terrain (slope), soils and drainage delimitations. Watershed is an appropriate unit for environmental (or ecological) planning for sustainable management of natural resources of a region. *Watershed Management* is a practice of conserving soil, water and biological resources using scientific principles, traditional and systems knowledge, and local resources with an objective of increasing crop productivity. It involves rational utilization of land and water resources for optimum production with minimum hazards to national resources. It essentially relates to soil and water conservation in the watershed which means land use according to land potential, protection of land against all kinds of deterioration, building and maintaining soil fertility, conserving water for farm use, proper management of water for drainage, flood protection, sediment reduction, and increasing productivity for all kinds of land uses (Tejwani, and Tideman as cited in Das, 1998).

Watershed Management in India

Watershed management has come into focus in India with the advent of productivity fluctuations with rainfall, necessitating micro-irrigation in drier parts, and also with the advent of space technology tools, which are useful in the micro-level planning. Land water related management projects and schemes have been implemented under various programs since the beginning of the Five Year plans. In particular, the Third Five Year plan introduced the watershed as the basic hydrological unit for soil conservation planning and execution (Iyengar, 1996). Increased emphasis on watershed development programs for dry land plain regions in India, *inter alia*, is a manifestation of the shifting priorities in the agricultural sector, which until recently concentrated mainly on crops and regions with assured irrigation (Shah, 1998). Successful case studies of Ralegaonsiddhi, Myrada, are well known (Singh, 1995).

An integrated watershed development approach was taken for sustainable management of natural resources with micro-level planning. The main components of watershed management programs undertaken were:

- Creation of a database for macro and micro level planning;
- Awareness program for policy makers, planners, soil conservationists, foresters, scientists and farmers and;
- People's participation.

The government had shown concerted efforts towards soil and water conservation during the Third Five Year plan. Besides numerous schemes and projects undertaken in that direction, there are three major programs that were implemented on a watershed basis: (1) Soil conservation in the catchments of River Valley Projects (RVP), (2) Integrated watershed management in the catchments of Flood Prone Rivers (FPR), and (3) National Watershed Development Projects for Rainfed Areas (NWDPRA). For implementation of watershed projects, the government has taken an approach of identifying critical watersheds after prioritizing watersheds on the basis of criticality of natural resources (in terms of degradation indices likes sediment yield, etc.) and other socio-economic conditions prevalent at country/ state/district level. Recently, Rama Mohan Rao et al. (1999) has made an attempt to classify soil and water conservation zones in India.

Production Systems Planning

In the sequence of evolution of natural resources management using the watershed approach for sustainability of these resources, thrust is on the productivity of natural systems. It is the productivity of natural systems that needs to be conserved through planning so that the needs of an increasing population are met and the threat to their renewability is thwarted. The watershed approach is an ideal approach to carry out a planning operation, and its planning framework shall fit well under the implementation and execution activities. Central to the success of the process is the participation of the population during the crucial implementation. Production systems planning (PSP) is a method of planning for the use of natural resources under the watershed approach with a focus on ecological characteristics. It essentially involves spatial allocation of land use for various production systems, namely, agriculture, horticulture, and animal husbandry, by which conservation goals are met through better decision-making. PSP is similar to regional land use planning, but it differs from it in that it depends on ecological characteristics at the watershed level rather than activities at the regional level. However, in both cases land use is an important element. The role of production systems in soil and water conservation is evident from the water and soil losses of catchments under such production systems such as mentioned by Mallik (as cited in Dhruvnarayan et al., 1989).

Table 1: Soil and Water Losses Under Various Production Systems

S. No.	Catchment Coverage	Loss of Water (cm)	Loss of Soil (kg/ha)
1.	Forest with normal cover	1	1
2.	Forest with poor ground cover	3	20
3.	Well managed pasture	3	14
4.	Agricultural crop	3	3250
5.	Grass lands (Pastures)	10	130

In most of the watershed management programs in India, emphasis is given on the operationalization of soil and water conservation program at the field level. PSP is a conservation-planning tool useful for watershed management through allocation of production systems to take care of soil and water conservation. PSP is not attempted in any of the watershed programs in India. However, the Indian Council for Agricultural Research (1973) has proposed a three-tier production system based on the terrain characteristics alone to serve as a guide for PSP.

Table 2: Terrain Based Production Systems Planning

S. No.	Terrain Category (Slope)	Production System
1.	Gently sloping to flat (< 5%)	Arable crops
2.	Flat to moderate sloping (5-30%)	Horticultural crops
3.	Steep (>30%)	Forestry

In the current study, we attempt to carry out the production systems planning based on the ecological characteristics of a watershed. A micro-watershed is demarcated from a larger watershed and natural characteristics of the ecosystems are captured under thematic categories of slope, soil, drainage and geomorphology. These along with the current land use are superimposed to prepare a production system plan. The methodology and procedure are described in the following section.

Case Study: Exercise of Production Systems Planning in Gutturu Watershed

Introduction

We chose for our case study a micro-watershed from the watershed priority map of the study district Anantapur in Andhra Pradesh (Andhra Pradesh State Remote Sensing Applications Centre, 1995). This drought prone district has precarious natural resources and is affected by the excessive use of environmental resources resulting in a qualitative and quantitative decline leading to the prevalence of desertification conditions (Ramakrishna, 1996). This watershed falls under the Penukonda administrative block of the district and has a major village named Gutturu, by which it is named.

Data Analysis

As a first step of the exercise, thematic maps of the watershed describing the ecological characteristics were prepared. The slope and drainage maps were prepared using the topographic maps of the watershed. The maps of geomorphology, soils and land use were prepared from visual interpretation of satellite imagery of IRS-1B provided by the Andhra Pradesh State Remote Sensing Applications Centre (APSRAC). The resulting classification under various themes and the maps created under each theme are presented hereunder. However, combining these themes into one single classification scheme is very difficult and often requires overlaying maps. The normal practice of analyzing overlain maps is to prepare a composite map of all classes and then to reclassify into broad categories so that a suggestive measure is derived using the field knowledge and experience of an expert. This is verified through field level examination.

Methodology

In the current study a little deviation is made such that the composite map is drawn from thematic maps after assigning weights to themes. The criterion for judgment of production system is based on the weighted thematic map of the watershed. The weights are given on the basis of importance of the theme's ecological function in the whole watershed and ranking of the themes based on importance. The weights assigned are subjective, but derived after consulting with experts in watershed management. Objective methods of weight assignment also exist in the literature (Ramanathan, 1994) but they are seldom applied in such problems. However, such methods, when applied, produce ideal results and help in finding near perfect solutions. Another approach to ranking themes is to conduct a Delphi survey of experts and then combine the opinions using numerical techniques. However, the method applied here is a beginning and it can be extended to make use of the techniques mentioned above as well as other advanced techniques in forming criterion for judgment. Further, the scope of systems planning can be extended when it is done at an even larger scale and incorporating socio-economic data on the watershed. The thematic maps of various classes for the chosen watershed are shown subsequent to the tables.

Table 3: Classification of Slope Theme

S. No.	Gradient (%)	Slope Class	Score	Weighted Score
1.	0-1	1	5	25
2.	1-3	2	5	25
3.	3-5	3	4	20
4.	5-8	4	4	20
5.	8-15	5	3	15
6.	15-35	6	3	15
7.	> 35	7	1	5

Table 4: Classification of Land Use Theme

S. No.	Land Use	Score	Weighted Score
1.	Double Cropped	5	20
2.	Kharif Rainfed	3	12
3.	Scrub Land	4	16
4.	Scrub Hill	4	16
5.	Barren Hill	1	4
6.	Fallow Land	2	8
7.	Saline Land	1	4
8.	Eroded Land	1	4

Table 5: Classification of Geomorphology Theme

S. No.	Map Unit	Geomorphic Unit	Groundwater Potential	Score	Weighted Score
1.	PPS	Shallow Weathered Pediplain	Poor to Moderate	2	6
2.	PPM	Moderately Weathered Pediplain	Moderate to Good	4	12
3.	PPMA	Moderately Weathered Pediplain Covered with Saline Soils	Moderate to Good	3	9
4.	P	Pediment	Poor	1	3
5.	DH	Denudational Hill	Poor	1	3
6.	RH	Residual Hill	Poor	1	3

Table 6: Classification of Drainage Theme

S. No.	Drainage Unit	Score	Weighted Score
1.	Dendritic	2	2
2.	Linear	1	1

Table 7: Classification of Soils Theme

S. No.	Soil Erosion Class	Score	Weighted Score
1.	Nil to Slight Erosion	2	4
2.	Slight to Moderate Erosion	1	2

Results

The aggregated scores are classified into 11 classes of various ranges mentioned in Table 8. For each category of the weighted score class, a suitable production system is allocated in consultation with a remote sensing and watershed expert. The production system plan for the watershed is prepared using this allocation method. The production system plan is shown for the watershed in the map following the table.

Table 8: Production System Allocation

S. No.	Weighted Aggregate Score	Planned Production System
1.	10-14	Forestry
2.	15-19	Forestry
3.	20-24	Silvipasture
4.	25-29	Fuel wood Plantation
5.	30-34	Social Forestry
6.	35-39	Conservation Agriculture
7.	40-44	Dryland Horticulture
8.	45-49	Dryland Agriculture
9.	50-54	Alley/Relay Cropping
10.	55-59	Intercropping
11.	>60	No intervention

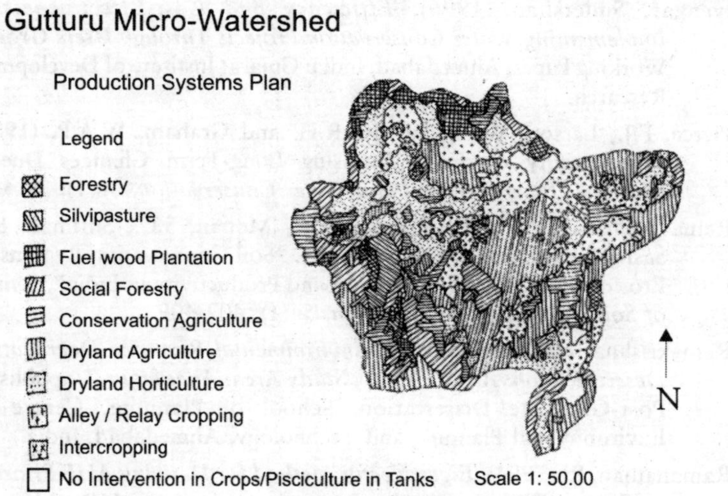

The production systems plan thus prepared based on the thematic maps and theme weights presents a methodologically different way of preparing watershed plan. The results can be directly translated into implementation with the help of wide range of stakeholders, i.e. public officials, non-governmental governments and people who are the landowners or managers in the watershed area with the agricultural farmers, horticulture farmers, or forestry personnel.

References

Andhra Pradesh State Remote Sensing Applications Centre (APSRAC) (1995). *Prioritization of Watersheds in Anantapur: Approach and Mapping Using Remote Sensing: Project Report.* Hyderabad: APSRAC.

Brundtland, Gro Harlem. (1989). *Our Common Future.* Delhi: Oxford University Press.

Das, S.N. (1998). "Watershed Development: Planning and Strategy", *Journal of Rural Development,* 17(3): 463-479.

Dhruvnarayan, et al. (1989). *Watershed Development.* Indian Council for Agricultural Research.

Hanumantha Rao, C.H., Ray, S.K. and Subbarao, K. (1988). *Unstable Agriculture and Droughts: Implications for Policy.* New Delhi: Vikas Publishing House.

Indian Council for Agricultural Research (1973). *Desertification and its Control.* New Delhi: ICAR.

Iyengar, Sudershan (1996). *Efficiency and Cost-Effectiveness in Implementing Water Conservation Projects Through Users Groups,* Working Paper. Ahmedabad, India: Gujarat Institute of Development Research.

Pierce, F.J., Larson, W.E., Dowdy, R.H. and Graham, W.A.P. (1983). "Productivity of Soils: Assessing Long-Term Changes Due to Erosion", *Journal of Soil and Water Conservation,* 38(1): 39-44.

Rama Mohan Rao, M.S., Sharda, V.N., Mohan, S.C., Shrimali, S.S., Sastry, G., Navain, P., et al. (1999). "Soil Conservation Regions for Erosion Control and Sustained Land Productivity in India", *Journal of Soil and Water Conservation,* 54(1): 402-409.

Ramakrishna, Nallathiga (1996). *Environmental Resource Degradation, Desertification: A Case Study, Study Area: Anantapur,* Unpublished Post-Graduate Dissertation, School of Planning, Centre for Environmental Planning and Technology, Ahmedabad, India.

Ramanathan, R. (1994). *Eigen Vector Method for Deriving AHP Priorities in System With Feedback,* (Discussion paper no. 119). Mumbai, India: Indira Gandhi Institute of Development Research.

Rosenberry, P., Knutson, R. and Harmon, L. (1980). "Predicting the Effects of Soil Depletion from Erosion", *Journal of Soil and Water Conservation,* 35(3): 131-134.

Shah, Amita (1998). *Watershed Development Programmes in India: Emerging Issues for Environment—Development Perspectives* (Working paper no. 96). Gota, Ahmedabad, India: Gujarat Institute of Development Research.

Singh, Katar (1995). *The Watershed Management Approach to Sustainability of Renewable Common Pool Natural Resources: Lessons of India's Experience*. Anand, India: Institute of Rural Management.

World Commission on Environment and Development (1987). *Our Common Future*. New Delhi: Oxford University Press.

Arcology and Arcosanti[1]: Towards a Sustainable Built Environment

10

David Grierson

Around the world, as cities reach unprecedented sizes, their increasing social and environmental problems need to be addressed if we are to avoid catastrophe. Paolo Soleri's arcology model aims at a more balanced relationship between urban form and efficiency of performance within a unique conception of the modern city. Since 1970 a prototype has been constructed at Arcosanti in the central Arizona desert to test the validity of the arcology model exploring such issues as the intensification in the use of space, higher residential densities, centralization, compactness, the integration of land uses, and self-containment of habitat. This essay describes both the arcology theory and the Arcosanti project and how the related ongoing work has wider significance in responding to some of the overlapping challenges that are involved in a movement towards more sustainable built environments.

"We do not inherit the earth from our parents, we borrow it from our children"

A Kenyan proverb

"...and we have no right, by any thing we do or neglect, to involve them in unnecessary penalties, or deprive them of benefits which it was in our power to bequeath."

John Ruskin, *The Seven Lamps of Architecture*

I

Over one hundred and fifty years ago John Ruskin, in his treatise on reforming architecture, described the earth as a "great entail", belonging as much to those who are to come after us, as to us. We had no right, he suggested, either through our action or inaction, to deprive our children of benefits that we might otherwise provide for them. In the *Lamp of Memory* he gave expression to a sensibility that now echoes around the world and has in recent years gathered momentum. Although there are many, often conflicting, definitions of what *sustainability* is, or might become, the move towards the construction a more sustainable society, if it is be achieved at all, will be founded upon a redefined relationship between the built and natural environments.

Throughout the 20th century the switch from a spread out pattern of human settlement to one of concentration in urban centers has spawned a built environment that has transformed the natural environment and contributed, in large part, to the current environmental crisis. Although the process of urbanization has occurred at varying rates throughout human history (Davis, 1965) the speed and nature of growth in cities in recent years, and the scale of environmental impact, is such that our own place in the history of the built environment is unique. Since the middle of the 20th century the population of the world's cities has soared from 200 million to almost 3 billion. This process is set to reach a significant stage in 2006, when for the first time in human history, one half of the global population (an estimated 3.3 billion) will live in and around cities (United Nations, Population Division, 1995). Although physically they occupy just 2% of the earth's surface, cities consume most of the world's natural resources, produce vast amounts of waste, and are the main source of pollution throughout the world.

In the developed world, urban environments consume so many resources that they are dramatically reducing the global resource base, or natural capital, on which we depend. They account for around 78% of the carbon emissions from human activities, 76% of industrial wood use, and 60% of the water tapped for use by people (Brown, 1999). In their current form, as cities grow they need to draw on more and more of the earth's capital. The ecological footprint (source of resources and sites of waste disposal and pollution) of cities like London, Tokyo and Paris extend far beyond their physical boundaries (Rees and Wackernagel, 1996). The amount of land needed to generate the resources to sustain the population of London (e.g. with food and timber), for example, is only slightly less than the entire land area of the United Kingdom (Girardet, 1996). In the developing world's fastest growing cities (Phoenix, Los Angeles and San Diego in the American Southwest) up to 85 metric tons of natural resources are consumed per person per year (World Resource Institute, 1996). This level of consumption represents a truly massive scale of environmental alteration. Extending this kind of resource intensive economic model to developing nations, as is now occurring around the world, is simply not environmentally sustainable.

Currently, on average, about three-quarters of the population within developed nations live in cities. Although some are still growing rapidly, this is mostly due to urban-to-urban migration. However, cities in undeveloped parts of the world are now growing five times faster. Over the next 20 years the fastest growth rate will occur in Asia and Africa. During this period Asia's urban population is expected to increase by a staggering 1 billion people (United Nations Development Programme, 1997). By then it is expected that there will be 33 mega-cities (a city with a population exceeding 8 million) in the world. Twenty-seven will have emerged within developing nations. Twelve will be in Asia. Only one—New York—will be in the United States. There will be none in Europe.

These cities are growing in number and size because they provide, generally, greater health, social and economic benefits than rural areas. But along with the benefits of urbanization have come environmental and social problems of staggering proportions. There is no historical precedent for changes of the scale and speed we are now witnessing and there is little guidance as to the magnitude of problems that such growth might pose. The environmental by-products of large and concentrated poor urban populations pose direct threats to health and to the quality of city life caused by problems such as air pollution and ozone build-up. Many people now live in life-threatening conditions. Today more than 600 million people in the undeveloped world are living in cities without adequate shelter. At least 220 million of the urban poor lack any access to clean drinking water and more than 420 million do not have the simplest latrines. Between one and two-thirds of the solid waste generated in these cities is left to pile up in streets and drains, contributing to flooding and the spread of disease, adding to an enormous toll of largely preventable deaths. In 1995 polluted air in 36 cities in India alone killed some 52,000 people. 1.1 billion people live in urban areas where air pollution exceeds health limits (Hardoy, Mitlin and Satterthwaite, 1992).

The transition to an urbanized world continues to have a profound effect on the physical arrangement of cities, their resource consumption, and their environmental impact. Throughout the last decade a cluster of United Nations conferences and conventions, beginning with the Earth Summit in 1992, have discussed the continuing deterioration of the ecosystems on which we depend for our well-being, and identified a common need to develop a global partnership for sustainable development. Recognizing that conditions in the world's cities, particularly those in undeveloped nations, had reached crisis point, the Habitat II (the 'City Summit') Conference, held in Istanbul in 1996, concluded with a global call to action to improve the quality of life within human settlements in cities around the world. Paragraph 4 of *The Habitat Agenda: Istanbul Declaration on Human Settlements* identified a need to confront problems such as:

- unsustainable consumption and production patterns, particularly in industrialized countries;
- unsustainable population changes, including changes in structure and distribution, giving priority consideration to the tendency towards excessive population concentration;
- homelessness;

- increasing poverty;
- unemployment;
- social exclusion;
- family instability;
- inadequate resources;
- lack of basic infrastructure and services;
- lack of adequate planning;
- growing insecurity and violence;
- environmental degradation; and
- increased vulnerability to disasters.

A pro-growth development paradigm has underpinned the formation of the built environment throughout the 20th century. The Club of Rome's *Limits to Growth* report (Meadows, Meadows, Randers and Behrens, 1972) highlighted an inherent fundamental flaw, pointing out that global growth in population numbers, resource use, waste production and pollution is exponential. While this kind of growth displays a gentle and gradual curve for a long time it rapidly shoots up in a very short period of time. What might seem like a manageable rate of resource use and waste disposal can quickly result in dangerously low levels of available resources and dangerously high levels of pollution. Environmentalists now question whether the rapid growth of cities in recent years can be sustained. They argue that in order that the quality of life of their inhabitants be maintained, let alone enhanced, the aggregate impact of cities on the environment—a product of the relationship between population, per capita consumption or economic activity, and energy/material flow per unit—must be radically reduced.

II

Sustainability involves a move from a current condition of unsustainable activity towards a process of improvement and increased quality. Essentially the term is used to indicate a change of attitude towards prioritizing ways of life that are in balance with the current renewable resources of the ecosystem and the biosphere. How we might better understand the nature of this balance and its relationship to the built and natural environments, thus helping to inform decision-making at various levels, is the subject of ongoing research. Although we are unclear about how much damage has already been inflicted on the biosphere a precautionary approach is proposed as a practical way forward. In the face of inherent uncertainty, risk is not an appropriate approach. Failure to maintain a viable biosphere will be catastrophic and irreversible (McDonach and Yaneske, 2002).

The widespread interest in theories, ethics and practice concerning sustainability indicates an increasing concern about the adverse impacts that conventional models of development have had on the environment, in both the developed and undeveloped parts of the world. Today, as urban environmental problems have been brought more sharply into focus (Commission of the European Communities, 1990), sustainable development is being described as a fundamental goal. The term has been used in recent years to catalyze debate concerning the relationship between economic growth and the natural resource base on which it depends. It was provided a global definition by the Brundtland Commission report (World Commission on Environment and Development, 1987) as development that meets the needs of the present without compromising the ability of future generations to meet their own needs. The definition implies that a balance can be found between how much development we are able to make while still preserving the environment to the extent that it can sustain an acceptable quality of human life in the future. It is not about the environment *per se* but involves a more complex, holistic, and systemic approach that mediates between social, economic and environmental processes and cuts across disciplines and subject areas. By suggesting that environmental protection and continuing economic growth can be seen as mutually compatible, it attempts to displace the *limits to growth* argument.

Although often criticized as being vague and even contradictory, the concept of sustainable development has, in recent years, begun to achieve political priority status among government institutions seeking a planned response to urban environmental problems. Alternative ways of re-thinking and reforming the built environment within a rapidly urbanizing world are now being considered. Within the academic and policy literature emerging around the notion of sustainable cities, a number of different models have been developed which represent different views of how such environments might be realized (Haughton and Hunter, 1994). Some advocate the re-designing the physical fabric of the city in order to improve resource efficiency and bring about more self-reliant settlements. This approach involves the frugal use of energy and natural resources while working within the carrying capacity of the local region. It also implies that the ecological footprint and energy budget for human settlements are vastly reduced. A move towards a smaller, more compact settlement pattern interspersed with productive areas for collecting energy, to grow crops for food, fiber and energy, and recycle wastes has been suggested as the way forward. The compact city idea is now being promoted as a major component of the various strategies emerging to tackle these problems (Breheny, 1992). The rationale for its implementation lies in a set of benefits that are seen as the outcome of more compact urban forms in which travel distances are reduced lessening fuel emissions, rural land is saved from development, local facilities are supported, and local areas become more autonomous.

III

The Italian born architect Paolo Soleri (1919 -) has dedicated much of his life to defining a different kind of urban environment. His model builds upon his work in the area of architecture and human ecology dating back to the 1940s. His thinking is holistic in the sense that it crosses traditional boundaries between subjects like architecture, ecology, biology, urban design, sociology, environmental studies and art. The scope of his early work was focused on the production of a theoretical model for a new physical landscape, what he calls a *neonature*, designed to support biological, human and social evolution while containing human societies along with all their material goods. The model stands in opposition to urban sprawl with its inherently wasteful consumption of land, energy, time and human resources, and advocates the need for a more balanced relationship between morphology and performance within a unique conception of the modern city. His approach recognizes the need for the radical reorganization of sprawl into dense, integrated compact urban structures in which material recycling, waste reduction, and the use of renewable energy sources are part of a strategy aimed at reducing the flow of resources and products through the urban system. Soleri refers to the new structure as an "arcology" (the fusion of architecture with ecology) to underline the conceptual basis, both in the discipline of architecture and the science of ecology. The concept was first published in the 1969 book *Arcology: The City in the Image of Man:*

> Such a structure would take the place of the natural landscape inasmuch as it would constitute the new topography to be dealt with. This man-made topography would differ from the natural topography in the following ways:
>
> - It would not be a one-surface configuration but a multilevel one.
> - It would be conceived in such a way as to be the carrier of all the elements that make the physical life of the city possible—places and inlets for people, freight, water, power, climate, telephone; places and outlets for people, freight, waste, mail, products, and so forth.
> - It would be a large-dimensioned sheltering device, fractioning three-dimensional space in large and small subspaces, making its own weather and its own cityscape.
> - It would be the major vessel for massive flow of people and things within and toward the outside of the city.
> - It would be the organizing pattern and anchorage for private and public institutions of the city.
> - It would be the focal structure for the complex and ever-changing life of the city.

- It would be the unmistakable expression of man the maker and the creator. It would be diverse and singular in all of its realizations. Arcology would be surrounded by an uncluttered, open landscape (Soleri, 1969, p. 13).

Soleri's model sees in the contraction and greater sophistication of the city (including all its equipment, machinery, infrastructure, services, etc.) both the efficient possibility of achieving more with less and the chance of reaching new levels of human development. A key function of the model is to facilitate a breakthrough to these new levels.

Soleri believes that the architect's task of ecological design should be directed towards the attainment of the progressive transformation of human existence. Inspired by the priest and paleontologist Pierre Teilhard de Chardin's (1881-1955) *point Omega* hypothesis (Teilhard de Chardin, 1959), Soleri believes that these environments can be instrumental in human evolutionary terms. He asserts that by adopting a more frugal lifestyle inside an arcology, its citizens would have the potential not only to do less harm to the planet but also to develop themselves spiritually (Soleri, 1981).

The approach clearly presupposes a radical revision of existing social, cultural, political and economic structures. Soleri points out the direction that must be taken within the *complexity-miniaturization-duration* (CMD) paradigm.

1. Complexity. Many events and processes cluster wherever a living process is going on. The make-up of the process is immensely complex and ever intensifying.
2. Miniaturization. The nature of complexity demands the rigorous utilization of all resources—mass-energy and space-time, for example. Therefore, whenever complexity is at work, miniaturization is mandated and a part of the process.
3. Duration. Process implies extension of time. Temporal extension is warped by living stuff into acts of duration, that is, the eventual "living outside of time" (Soleri, 1983, p. 17).

Since all of nature complies with an imperative involving these three fundamental principles, Soleri argues that the human habitat must be also designed to conform. Sustainability, in arcological terms, thus becomes part of an evolutionary process. He points out that in nature, as an organism evolves, it becomes more complex and tends towards a more compact and miniaturized form, or system. Since more events can occur in a more complex system he argues that successful and sophisticated (sustainable) human settlements must follow the complexity/miniaturization/duration path in order to become more lively containers for social, cultural and spiritual development (Soleri, 1973).

Soleri's methodology was developed within the Mesa City Project (1958-67) where a series of relationships between human ecology and urban design were explored.[2] In 1969 he published *Arcology: The City in the Image of Man* and sketched out giant structures that would dwarf the Empire State Building and were located in the most inhospitable environments imaginable. The 31st generation arcologies had names like *Novanoah II* (a city for 2,400,000 to float on coastal waters or open sea), *Stonebow* (200,000 people above a ravine or canyon), *Theology* (with a population of 13,000 set within a cliff), *Arcube* (a city of 400,000 people located on flatlands), *Hexahedron* (a city of 170,000 on any topography), and *Asteromo* (70,000 people living in space).

Later, in response to the growing energy crisis of the mid-70s, the architectural concept of the first generation structures was split in half exposing the core of the structure to the sun. This next generation of arcologies described some highly significant related effects; it gave greater priority to the main source of renewable energy and placed the arcology model much more firmly into its own ecological niche. It also served to reinforce the relationship between architecture and ecology through the concept of the *energy-city* wherein the entire settlement is conceived as an instrument for energy conservation and environmental sustainability, to be highly efficient in the processes of production and consumption. Six major (and simple) architectural *effects* were described, collectively under the "Two Suns Arcology":

> The *greenhouse effect* is a membrane that seals off an area of ground that can be cultivated, extending the growing season to practically twelve months, and also saves a great amount of water.... With the "greenhouse," one has intensive agriculture, limited use of water and extension of seasonal cycles. This is the *horticultural effect.* Then there is the *apse effect.* Some structures can take in the benign radiation of the sun in winter months, and tend to cut off the harsh radiation of the sun in the summer. By the *chimney effect,* which is connected with the greenhouse effect, one can convey, passively, energy through the movement of air and heat from one area to another. So we have these four effects; there is also the capacity of masonry to accumulate and store energy—*the heat sink effect.* With relatively large masonry, one can store energy during the warm hours of the day, and give it out during cool or cold hours of the night. The intent is to see if these five effects can be organized around what I call the *urban effect.*[3] The urban effect is the capacity of mineral matter, to become lively, sensitive, responsive, memorizing. ...If we were to co-ordinate those six effects together, then we definitely could save on resources like land, water, time, energy, materialism, and have a better ecological sanity (Soleri, 1985, p. 137).

These effects were combined in a series of designs for second-generation arcologies, including *AirDam*, *India Village*, *Maryland* and *Regina.* Within these proposals the entire form of the urban structure as well as a huge area of south-facing greenhouses containing vegetable gardens, are designed to maximize the use of solar energy while reducing dependence on external energy sources.

IV

Within a second generation arcology, material recycling, waste reduction, energy conservation, and the use of renewable energy sources, like sun and wind power, would offer the basis of a strategy for sustainability that aims towards a more efficient process of urban production and consumption (Grierson, 2000). The drawing together of diverse city functions into mixed-use, self-contained arcologies would encourage cultural intensification and social integration within their boundaries, while freeing up the surrounding hinterland to remain natural.

We can see within the *Two Suns* approach a theoretical response capable of confronting many of today's urban environmental problems. These include the inefficient use of land, air and water; pollution caused by technological society; energy and natural resource depletion; the distribution and consumption of resources; food scarcity; the loss of quality of life through waste; the affluence and opulence and the physical and social segregation of people, things and activities; and the increasing problems of social alienation and exclusion. These are among the key issues that have begun to concern contemporary urban planners and architects working on urban sustainability, such as Peter Calthorpe (1993), Herbert Girardet (1992), Richard Register (1987) and Richard Rogers (1997). What sets Soleri apart is that he has been proposing a methodology for an ecological model in urban design for more than 50 years and at the same time trying to illustrate it through the arcology projects. Perhaps more significantly, over the last four decades he has been attempting to build a prototype in the semi-arid desert region of central Arizona in the United States to test his ideas. Since construction work began in 1970, Soleri, with the help of a dedicated community of volunteers, has designed, built and inhabited Arcosanti (meaning *architecture before things*) as an urban laboratory. The laboratory attempts to confront problems associated with:

- population expansion and land use;
- social integration;
- self-containment of habitat;
- urban transportation;
- food and energy production; and
- the habitat's impact on natural resources and pollution.

Soleri argues that our ecological sanity is dependent on centers of life so intense as to contain the vast bulk of the planet's population and all its paraphernalia. Through an adherence to the CMD paradigm, arcology is therefore dedicated to the "old" notion of containment as opposed to the relatively recent phenomenon of (a car-driven) diaspora. The "City in the Image of Man" metaphor articulates this idea of self-containment. Without self-containment the city cannot act effectively with the surrounding natural environment. In contrast to sprawl, Arcosanti offers a working prototype for a new kind of city that is designed, built, and inhabited as a three-dimensional, highly concentrated urban structure. Its compact design allows agricultural land and biologically diverse habitats to remain preserved beyond the city's perimeter. A permanent experiment in urban intensity, when complete Arcosanti will house an environmentally benign community of five to six thousand people on only 15 acres of land—equating to a population density of around 350 people per acre, 10 times the population density of New York City.

Set on the edge of a mesa above the Agua Fria River in the middle of an 860-acre nature preserve containing orchards, agricultural fields, canyons, and high desert hills, the compact structures face toward the sun to gather its energy. When complete they will stretch no more than quarter of a mile on any one side but will rise to as much as thirty stories. Inside, these structures will contain the economic, cultural and social infrastructure normally scattered around a modern city, while providing residents up to two thousand square feet of living space per family. A series of orchards will line the north side of the structure, creating a unique fusion of urban and agricultural environments. Outside there will be expansive views of another three thousand acres leased from the state of Arizona, to be kept as undeveloped open space. An integral part of the design will be five to seven acres of south facing sloping greenhouses, an "energy apron" acting as a central system for producing food and collecting energy to support the prototype town.

Social integration is a main goal at Arcosanti and points to the reshaping of the entire urban landscape and along with it the culture that such a landscape supports. Our information age offers society an unprecedented opportunity to bring together the main components of life, but the habitat that we have constructed for ourselves during the last century is alien to such integration. Therefore, Soleri argues, it needs to be reconfigured. Arcology advocates mixed use in its purest form—accommodating a variety of uses within one structure. In an attempt to reintegrate people within their community, Arcosanti is designed as a mixed-use complex, containing homes, offices, schools, parks and a cultural center. The belief is that a close interaction of city functions and people will induce a greater sense of community. The integration of living, working and recreation become a very natural part of things at Arcosanti. It is something that is to be experienced in everyday life.

Many civilizations throughout history have had to deal with restrictive ecosystems. The 1969 arcologies were located on marginal lands, far from main transportation networks, many poor in resources. Since these may be the sort of reserves where future cultures will have to settle (leaving fertile lands free for increasing crop cultivation) the task has been to demonstrate the viability of the self-containment of a community on such inhospitable land. Arcosanti's semi-arid location in the desert presents particular challenges to settlement but by adopting a higher concentration of land use deriving from a mixed use development it demonstrates an effective method of altering the impact of a settlement on the natural environment.

Rather than accept the logic of two-dimensional cities, since 1970 Soleri has used Arcosanti to rethink modern urban planning and explore the idea of *urban implosion*—wherein the habitat's infrastructure contracts and intensifies in order to become more efficient, ecological and sustainable. The ecological significance of the first generation arcologies derives mainly from the elimination of the motorcar, and the reclamation of all the space that is normally associated with this form of transportation. Inside an arcology walking would be the main form of transportation. Typically today's cities devote up to 60% of their land for car functions such as roads, car parks, showrooms, garages, petrol stations, repairs and junkyards. Eliminating the car from inside the arcology and reserving it for use outside dramatically reduce that figure leaving the land free for other use. The unique character of Arcosanti will derive, in large part, from the prohibition of the motorcar and the widespread use of pedestrian walkways, lifts, escalators and moving platforms. Because of its compact nature most journeys by foot will take about 15 to 20 minutes (about the same time as it takes, typically, to walk from inside a shopping mall to the outskirts of the car park in cities like Phoenix and Los Angeles). Like many critics of car dependency Soleri is not against the technology *per se* but against our complete reliance on it for transportation. While he acknowledges planning efforts that aim to produce more efficient land use patterns thereby reducing the number and frequency of car trips, and the introduction of fuel-efficient technology in car design and manufacture, Soleri contends that such improvements fail to attack the core of the problem. They are, he says, simply "a better kind of wrongness". By virtue of its compact design, Arcosanti would allow cars to be relegated to service areas on the periphery or reserved for travel between communities.

Arcology aims at a degree of autonomy and self-reliance, rather than self-sufficiency. Self-sufficient communities, which aim at total self-provision of food and energy and the complete recycling of wastes, are, according to Soleri, "extravagant and devoid of sense". There is no way the earth or anything in it, he argues, can be perfect because it is a small part of a much larger system. The arcology concept is directed instead at a more restrained and judicious use of resources via the power of complexity and miniaturization and the discipline of frugality. The degree of self-reliance in food production has changed as the arcology concept has evolved. In *Mesa City* the settlement was designed to be entirely dependent on the produce of the surrounding hinterland and on traditional agricultural practices. With the development of the *Two Suns* approach food and radiant energy are produced within south-facing greenhouses located within the city. These are designed to support the city's population at a minimum level. Other produce is imported from outside to supplement the goods and services provided on this self-reliant base (e.g. electricity from the main grid). All Soleri's projects after 1958 have explored methods of generating and harvesting energy from renewable sources and have aimed at transforming the urban structure into an *energy machine*. In the *Two Suns* approach the city is conceived of as a complex in which living, working and learning are integrated with food and energy production. The city becomes both consumer and producer. *Apses* and *exedra* (semi-circular edifices—developed from the apse form) that respond to the Sun's trajectory as energy devices and large expanses of greenhouses are attached to the city, and used to generate heat and electricity as well as to grow food. These help define the urban structure as an *energy city*.

Arcologies, through a blend of energy conservation and land use efficiencies together with waste recycling systems, could maintain the ecological integrity of the region while placing fewer demands on the environment in terms of land, water, soil, fuel and other resources. By reducing the demand for petrol-based transportation systems, air quality could be radically improved. Non-pollutant, passive solar energy systems such as wind turbines, photovoltaic cells, and solar cooling and heating would further help to reduce water, air and land pollution. By combining various passive energy strategies within a single integrated urban system, Arcosanti aims at a theoretical and architectural synthesis in which philosophical, ecological, and theological ideas are "woven into the physical structure of great beauty and integrity and which, at the same time, is a structure of stunning frugality from the standpoint of energy conservation" (Skolimowski, 1975, p. 33).

V

While following in the tradition of the 1960's libertarian counter-culture communes and the more recent eco-villages movement, the community at Arcosanti is unique in that its motivation lies primarily in the production of ecological architecture. The main focus of the community has been in building the project's various concrete structures using an earth-casting construction method based on ancient craft techniques. These structures now define the living, working and learning processes of around one hundred residents. As well as ongoing construction work, residents are involved in a variety of activities ranging from project development, research activity, exhibitions, drafting and design work, conference organization, to wind-bell production and site maintenance.

Arcosanti is now officially listed on the state map of Arizona as a small town. As such it is faced with the challenges of daily existence but, at the same time, its aim is fixed firmly on the future. By trying to anticipate it, and moreover attempting to plan for it, the community strives to keep the road to the future open while recognizing that paradise here on earth can only ever be an imaginary condition. With each passing year Phoenix creeps ever closer to Arcosanti. It is conceivable that, in the not-too-distant future, Arcosanti will be another of its suburbs.

Endnotes

1. *Arcology* is a fused word adopted by Paolo Soleri to describe both a theory of urban design rooted in the discipline of architecture and the science of ecology as well as a physical structure designed according to arcological principles. Arcosanti is the name given by Soleri to the 30th arcology published in 1969 in *Arcology: The City in the Image of Man*. Arcosanti was designed to be an experimental complex providing a testing ground for arcological concepts. Like *arcology*, *arcosanti* is a fused word of Italian origin. *Arco* refers to architecture (or the arch), *cosa* is a widely used word meaning *thing*, and the annex *nti* suggests a certain permanence or timeless quality. Together the name suggests a non-material (or spiritual) process at the heart of architecture—"architecture before things".
2. Mesa City was a theoretical regional plan to house 2 million people on around 55,000 acres (about the size of Manhattan Island) on an isolated and pre-flattened desert plateau. Developed as part of Soleri's ongoing research in the field of "architecture as human ecology", Mesa City was specifically aimed at introducing the idea of "corposity into the urban morphology, a premonition of the arcological concept" (Soleri, 1969).
3. The *urban effect* is described as a universal effect involving the transformation of mineral matter into mind via the potentially unlimited power of complexification and miniaturization. Soleri says it is that fundamental phenomenon in which two or more particles of physical matter begin to interact in ways other than statistical or fatal, that is, in ways that are organic or living.

References

Breheny, M. (ed.). (1992). *Sustainable Development and Urban Form.* London: Pion.

Brown, L., et al. (1999). "Winning or Losing the Environmental Battle: Cities Hold the Key to Planetary Health" [Press release]. Washington, DC: The World Watch Institute.

Calthorpe, P. (1993). *The Next American Metropolis: Ecology, Community and the American Dream.* New York: Princeton Architectural Press.

Commission of the European Communities (CEC). (1990). *The Green Paper on the Urban Environment: Communication from the Commission to the Council and Parliament.* Luxembourg: Office for Official Publications of the European Community.

Davis, K. (1965). "The Urbanization of the Human Population", *Scientific American,* 213(3): 41-53.

Girardet, H. (1996). *The Gaia Atlas of Cities.* London: Gaia Books Ltd.

Grierson, D. (2000). *Ecology, Sustainability and the City: Towards an Ecological Approach to Environmental Sustainability: With a Case Study on Arcosanti in Arizona.* Unpublished doctoral dissertation, University of Strathclyde, Glasgow, Scotland.

Hardoy, J.E., Mitlin, D. and Satterthwaite, D. (1992). *Environmental Problems in Third World Cities.* London: Earthscan.

Haughton, G. and Hunter, C. (1994). *Sustainable Cities.* London: Jessica Kingsley Regional Studies Association.

McDonach, K. and Yaneske, P. (2002). "Environmental Management Systems and Sustainable Development", *The Environmentalist,* 22(3): 217-226.

Meadows, D.H., Meadows, D.L., Randers, J. and Behrens, W.W. III. (1972). *The Limits to Growth: A Report to the Club of Rome's Project on the Predicament of Mankind.* New York: Universe Books.

Register, R. (1987). *Ecocity Berkeley: Building Cities for a Healthy Future.* Berkeley, CA: North Atlantic Books.

Rees, W.E. and Wackernagel, M. (1996). *Our Ecological Footprint: Reducing Human Impact on Earth.* Philadelphia: New Society.

Rogers, R. (1997). *Cities for a Small Planet.* London: Faber and Faber.

Skolimowski, H. (1975). "A Comment on the Two Suns Arcology", *aaq: Architectural Association Quarterly,* 7(2): 33.

Soleri, P. (1969). *Arcology: The City in the Image of Man.* Cambridge, MA: MIT Press.

Soleri, P. (1973). *The Bridge Between Matter and Spirit is Matter Becoming Spirit.* Garden City, NY: Anchor Press/Doubleday.

Soleri, P. (1981). *The Omega Seed: An Eschatological Hypothesis.* Garden City, NY: Anchor Press/Doubleday.

Soleri, P. (1983). *Arcosanti: An Urban Laboratory?* San Diego, CA: Avant Books and the Cosanti Foundation.

Soleri, P. (1985), *Technology and Cosmogenesis.* New York: Paragon House.

Teilhard de Chardin, P. (1959). *The Phenomenon of Man.* London, UK: Collins.

United Nations Development Programme (1997). *Human Development Report 1997.* New York: UNDP.

United Nations Population Division (1995). *World Urbanization Prospects: The 1994 Revision.* New York: United Nations, Dept. for Economic and Social Information and Policy Analysis, Population Division.

World Commission on Environment and Development (1987). *Our Common Future.* Oxford: Oxford University Press.

World Resource Institute (1996). *World Resources: A Guide to the Global Environment.* New York: Oxford University Press.

Unheard Voices: Environmental Equity

11

Rosemarie Russo

A fairly large and growing body of literature today presupposes that there are substantial public health costs borne primarily by minority, low-income, and other disadvantaged populations specifically because of their differential exposure to environmental hazards (Bowen, 2001). As a democratic society, we must strive to set an example and improve the conditions of those vulnerable to the forces of inequitable power relations and social systems. Environmental justice principles seek to: affirm the right to protection, prevent harm, shift the burden of proof, obviate the requirement to provide proof of intent to discriminate, and target resources to redress inequities (Bullard, 1993). Environmental equity is premised on the notion of fairness in the distribution of environmental risks, particularly those of a technological origin (Tarlock, 1994). Environmental equity is complex because it encompasses scientific studies, judicial decisions, social policy, international trade, ethics, and conflicting values. This article briefly describes environmental justice issues facing urban and rural communities in the United States and address one particular problem—lead poisoning—since it has become apparent that the incidence of subclinical lead poisoning is overwhelming high among minority populations in low-income urban areas (Giegengack, Cressler and Bloch, 1999). It also looks at the connections between racism, sexism, cultural destruction, power relations, and environmental equity.

Historically, environmental justice was viewed as a movement to prevent people of color from becoming victims of industrial pollution. The principles of the environmental justice movement in the United States were articulated by the delegates to the First National People of Color Environmental Leadership Summit (United Church of Christ Commission for Racial Justice, 1992). Simultaneously, community groups gathered strength across the nation, from the Mothers of East Los Angeles, formed by Juana Gutienez, who fought against a hazardous waste incinerator; to the Good Road coalition in Rosebud, South Dakota; all the way to the Evanston community in Cincinnati, Ohio. The Evanston community (population 8,300, with 95% people of color) formed a community center in response to three acts of environmental racism including the siting of a BASF storage and waste facility and the clean-up after an explosion at the facility which they contend occurred because of irresponsible management (Phillips, 1995). There is no buffer zone between the BASF facility and Evanston, while there is a buffer zone between the facility and the predominantly white communities of Norwood and Walnut Creek. Carol Browner, former administrator of the Environmental Protection Agency (EPA), confirmed that some communities do bear a disproportionate share of contamination problems (Environmental Protection Agency, 1994). It has become apparent that the disenfranchisement of minorities is strengthening an alliance of civil right advocates and environmentalists. Women in ecodevelopment have added the issue of sexism into the environmental justice arena. Thus, environmental justice goes beyond our borders.

As Garret Hardin (1968) discussed in *The Tragedy of the Commons,* resource exploitation leads to ecological, social, and institutional crises, and now such exploitation is on global scale that encompasses urban relocation issues in areas such as Africville, Nova Scotia, to environmental sustainability needs of women in highly populated regions of the world such as Africa and India. Environmental justice encompasses culture, race, gender, age, class, and power relations in issues ranging from health-related agricultural issues to inner-city toxic contamination of children. On a global scale, the World Health Organization estimates that 40,000 people die each year of pesticide poisoning, mostly in developing nations. However, California agriculture uses 10% of all pesticides in the world, ranking only 49th in sustainable farming practices (Jenks-Jay, 1999).

In this article, a main focus is to examine the common denominators among the exploitation of marginalized citizens in the communities of the United States and the less developed countries. One major commonality is the unheard voices in environmental decision-making. Within the United States this is partly because those citizens do not understand how to impact the legal process. In this article I provide references and strategies on how individuals and environmental groups can access information, gain legitimacy, and ameliorate environmental inequalities through policy and administrative responses. Without knowledge about the process it is like playing a game without the rulebook—one has to "trust" the referee. Evidence reveals that people of color are subjected to a disproportionately large number of health and environmental risks in their neighborhoods (e.g. childhood lead poisoning) and on their jobs (e.g. pesticide poisoning of farm workers) (Bryant and Mohai, 1992, p. 10). As an educator and legal analyst, my intention is to raise awareness about different tactics that can be implemented to revitalize communities locally and globally, since the present legal and regulatory system is failing certain segments of society. I hope to put local and national struggles into a global context to illustrate how the same tactics can be used to impact environmental challenges at all levels. The simple acknowledgement of others' living conditions and cultures within our own society is an excellent first step.

Grassroots leaders (many of them women) have emerged from groups of concerned citizens who see their families, homes, and communities threatened by some type of polluting industry or government policy (Bullard, 1990; Gibbs, 1982; Shiva, 1989; Pardo, 1990; Hamilton, 1990). Often times the issue of racial bias, or *perceived* racial bias, tends to create adversary situations that do not separate the people from the problem. I contend that to resolve environmental disputes, one needs to re-evaluate the complex cultural, scientific, economic, legal, and power issues surrounding environmental disputes.

The Game of Environmentalism

Environmentalism is like a chess game with different players, numerous strategies, and a racial dichotomy. It is a universal game that can be played by people that do not share a common language—like global environmental issues that demand moves from people that do not share a common language, culture, or ideological views. The board represents a fixed place—Mother Earth. The white players represent: multinationals, developed countries, courts, regulators, the upper class, traditional environmental groups, and politicians. The black players represent developing countries, women, community groups, and poor workers who must choose jobs over environmental concerns. Like a chess game, there is a temporal element to the environmental movement. Do we speed up the sixth mega extinction? Do we destroy the planet through nuclear annihilation in an instant? Or do we slowly cause irreversible ecological decay to the ecosystem through ozone depletion. Like chess, environmental decisions are often complex and require comprehensive strategies similar to chess. No one knows the outcome; each player has a different role. These are the players:

White

Pawns. Including the amateur environmentalist (who canvases but doesn't know all the issues); the recycler (limited effort but philosophically feels better); Jacques Costeau (role models who expand our comprehension of species and environments); the Sierra Club (organizations involved in land preservation—the stereotypical environmental movement types); the righteous environmentalist (Not in my Backyard—NIMBY advocates that drive gas guzzling SUVs to public meetings. Sport utility vehicles have remained one of the most popular vehicles among middle-class families in spite of the fact that baseline fuel efficiency is only 15.3 mpg (Wipke, 1998)); the activist of color who refuses to say "not in my backyard" without questioning or caring about whose backyard the problem ended up in (Bullard, 1993).

Rooks. A quite powerful piece, but often misplayed and kept out of the game too long for maximum benefit. Rooks include the safety manager/consultant and the educator who tries to raise awareness.

Bishops. Including the Supreme Court—the final authority on legal environmental cases, through cases such as *Sierra Club v. Costle* they dictate an individual's or entity's ability to affect proposed rules—and the regulating agency—the translators of lofty legislative goals such as "restoring the integrity of the nations water" language from the Clean Water Act. The EPA and state agencies are charged with issuing discharge permits for municipalities and industry based upon a water body's classification. These permits known as National Pollutant Discharge Elimination System (NPDES) permits are public records therefore citizens can obtain copies through the Freedom of Information Act.

Knights. Including multinationals such as the World Bank. These giant funding organizations that create large dams, like the ones underway on the Yangtze and Namada rivers, with little regard for ecological damage and human displacement. At stake also is the cultural loss in such places as the ancient city of Zeugma—a Byzantine city inundated by the Birecek Dam. Another noted example is the series of dams built on the Awash River. The World Bank funded the project, which provides water to the sugarcane, cotton, and banana plantations owned by wealthy Ethiopians and Dutch, Italian, Israeli, and British firms. The project effectively dried up the lands downstream and flooded lands upstream, uprooting more than 20,000 people (Shiva, 1989).

Queen. Rachel Carson, scholar, scientist, and activist, did more through her book *Silent Spring,* published in 1962, to force legislation such as the Federal Insecticide, Fungicide and Rodenticide Act (FIFRA) in spite of criticism by her peers. This queen is a symbol of interdisciplinary success.

King. The American consumer—the rich, white consumer who uses a disproportionate amount of energy. The average person in an industrial market economy uses more than 80 times as much energy as someone in sub-Saharan Africa (World Commission on Environment and Development, 1987).

Black

Pawns. Including the Maquiladora workers (working in factories that create pollution in the communities); Latino farm workers (exposing themselves and future generations of children to pesticides); lead-exposed African American children (urban poor living in public housing projects); urban residents (toxic donut residents, so called because they are surrounded by toxic sites); indigenous societies (such as the Hopis); fishermen (citizens that live off the land and are connected to their food source); hunters (contributors to Ducks Unlimited); and the politicians—representatives that introduce environmental legislation for campaign purposes. However, of the hundreds of bills that are introduced few are passed and even less are funded which make them little more than political propaganda. Take for example, Wayne Allard, Senator from Colorado, who was challenged in the November 2002 election by the League of Conservation Voters to verify his environmental campaign claims based on his dismal voting record that included voting to fence off rather than clean-up the Shattuck hazardous waste site, voting against funding for national parks, and voting for the roll-backs of the Clean Air Act's New Source Performance Review standards. In fact, Senator James Inholfe's bill to allow roll-backs was endorsed by the majority of the senators, while Senator Edwards' bill to delay roll-backs pending further study was defeated 46-50 (Environmental Defense Fund, 2003).

Rooks. Representing "soldiers" (paid players destroying the environment for the king). For example, the Forest Service has offered below-cost timber sales, in which the price paid by the loggers for the timber is lower than the costs incurred by the Service, to make the timber available. Below-cost timber sales resulted in average annual losses by the Forest Service (and ultimately by the taxpayers) of $ 327 million between 1979 and 1990 (Kaufman and Franz, 2000).

Bishops. Representing the United Church of Christ Commission on Racial Justice who challenged the EPA's policies on environmental equity (United Church of Christ Commission on Racial Justice, 1987), and Lois Gibbs and the Citizens Clearinghouse of Hazardous Waste, a database for citizens to learn about chemicals in their neighborhoods.

Knights. Include the First Nation People of Color Environmental Leadership Summit who commissioned a national conference on environmental justice, and the Environmental Defense Fund who represents communities and groups in environmental litigation and negotiations.

Queen. Vandana Shiva, physicist and author, representing ecofeminists from developing nations.

King. The Animal Kingdom—in all its diversity, from ciliates and amoebas to Saepiolus Blue butterflies.

King

The King moves only one square at a time. The temporal element of evolution. The king also represents the distinct lineages of species.

Queen

The Queen is the most powerful piece. She moves any number of squares in any direction. Like ecofeminists, self-proclaimed or not, the Queen's moves represent the distances that women have to take to obtain water and firewood. For example, in parts of India women have to walk 15 to 20 miles to fetch water (Shiva, 1984). The Queen can move in any direction—choosing to have no children or numerous children, thus controlling population. This alone makes her the most powerful environmentalist and player.

Bishops

These players move diagonally any numbers of squares forward or backward. Politicians and religious/community leaders represent Bishops. The politicians are influenced by special interests. For example, if a politician received campaign funds from a timber company, he might be less likely to vote for wilderness designation of an area that would not allow timber sales. The religious leaders move in different directions as well. Some religious leaders in Christian and Jewish traditions interpret and preach from the Old Testament that human divine rights include control over nature. Other religious societies, such as the Hopi, are devoted to securing blessings from nature. To the Hopi, objects such as springs are objects of religious veneration.

Rooks

The rooks can move any number of squares horizontally or vertically. The rook represents the environmental health and safety officer or consultant. He can influence a number of people or processes but has limited power (i.e. staff, money). The rook can also be a soldier representing the white King's interest at the expense of the environment.

Knights

The knights are the only pieces able to jump over another piece. The white knights are represented by multinationals that hold themselves as the saviors to developing countries in terms of providing capital for major infrastructures. Yet these projects, such as dams, literally and figuratively drown cultures and civilizations. The multinationals jump over the process of public participation. Our black knights are represented by the First National People of Color Environmental Leadership Summit and by coalitions of environmental and social justice groups, such as the National Resource Defense Council and the NAACP Legal Defense and Education Fund, who joined forces and won an out-of-court settlement of approximately $ 16 million for a program in California to test blood lead levels. *Matthews v. Coye* arose because the state was not performing the federally mandated lead testing of some 557,000 poor children who could pay for the testing through their Medicaid benefits but were not receiving the tests regardless of the fact that they were financially entitled to care (Westra and Wentz, 1995).

Pawns

A pawn can be moved either one or two squares forward on its first move. One of the pawns is the college student, moving with conviction and vigor initially, yet by the second decade his convictions are challenged as he is asked by his employer to represent anti-environmental clients. Moving more slowly, he becomes an armchair environmentalist. Another pawn does reach the opposite side and is promoted to queen.

Check and Checkmate

The environmental movement is similar to chess in the aspect that scientists warn us, just as a player announces, "check". The ending depends on the players and how they move—hopefully it is not a checkmate or perpetual check. And so we can come together to engage one another thoughtfully, or we can end the game in one quick deliberate move. The environmental justice movement promotes the concept of engaging one another in dialogue and has several important players, like Robert Bullard who claims environmental justice is a movement to prevent certain communities from becoming ecological "sacrifice zones". Beyond our border is ecofeminist Vandana Shiva, who believes environmental justice represents not only prejudices against nature but against women as well. Like the definition of environmentalism, environmental justice has a broad range of meanings. The chess analogy reflects the pieces, strategies, the complexity of environmental gambles, and the interrelationships between players, resources, the environment and development. The next sections provide examples of the interconnections between powerful players and affected members of society and strategies to balance the game.

Exporting Environmental Racism

An international perspective of environmental justice includes oppressors disguised as economic agencies, such as the Export-Import (Ex-Im) Bank of the United States, multinationals, and the World Bank. The Ex-Im, a government agency that supports U.S. multinationals through subsidized loans, has given a preliminary commitment toward a $ 100 million loan guarantee to the British engineering firm Balfour Beatty to support the Ilisu hydroelectric project in Turkey. This project will destroy the homes of more than 78,000 people, mostly Kurdish refugees. Although the project will produce an environmental impact statement it is unlikely there will be public input since the Kurdish population is afraid to speak out against the government (Knight, 2000).

In addition to eco-racism in relation to future development, less financially sound countries bear the burden of exposure from past industrial activities. On December 12, 1991, Lawrence Summers, chief economist of the World Bank, argued in a memo to colleagues that toxic waste storage and disposal should be located in poor countries (Westra and Wenz, 1995). This is based on the theory that the measurement of the costs of health-impairing pollution depends on the foregone earnings from increased morbidity and mortality (Westra, 1993).

As advocates in the United States focus on our rights to a safe environment, we may be exporting environmental racism abroad through the sale of banned pesticides or toxic waste disposal exports (EPA, 2003). The United States has yet to sign the international treaty on Persistent Organic Pollutants (POPs) and allows the export of chemicals covered in the treaty including but not limited to chlordane, dieldrin, DDT, heptachlor, hexachlorobenzene and toxaphene. Worldwide controls on POPs are essential because POPs migrate through the atmosphere and accumulate in the food chain. Despite the fact that most people believe that the produce they buy meets pesticide safety standards, a recent study that analyzed 14,923 computerized records from the Food and Drug Administration showed that pesticide violations are grossly underreported (Houlihan, Campbell and Wiles, 2000). During the study period one-quarter of all the green peas examined contained illegal pesticides, as did 15.7% of the pears, 12.5% of the apple juice, 11.7% of all the green onions, 7.6% of the green beans, and 7.4% of the strawberries. The high violation rates from major suppliers, such as a 40.8% violation rate on green peas from Guatemala and 18.4% violation rate on strawberries from Mexico, may be partially due to the fact that the United States provides these countries with instructions on acceptable use for pesticides in English only. Furthermore, although inspections for harmful pesticide residues take place pursuant to the Food, Drug and Cosmetic Act, a government accounting agency report says less than 1% of the imports are actually inspected (Firestone and Reed, 1993). If the food reaching U.S. borders contains these levels of pesticides, what are the levels within the importing countries? Are we exporting racism? Annual pesticide poisonings are said to be at least 1 million per year with upwards of 11,000 deaths each year worldwide.

As Rev. Benjamin Chavis (1993) explains in "Toxic Waste and Race", racism is more than just a personal attitude; it is the institutional form of that attitude. Both consciously and unconsciously, racism is enforced and maintained by the legal, cultural, religious, educational, and economic environmental institutions of societies. Many Africans have been encouraged to engage in a variety of export-driven economic activities that ultimately are environmentally detrimental. These activities include raising beef for exports, especially to European Union countries, on marginal land as in Botswana—an activity that also contributes to the expansion and deepening of the Kalahari Desert. And yet only a very tiny fraction of the population of Botswana benefits from beef exports (Kokole, 1995). Racism also exists in regards to imports. Ali Mahdi, a minister in Somali, signed a 20-year contract to accept toxic waste from Italy. During the second year of the contract a warehouse containing 81,200 liters of toxic chemical pesticides caught fire. The damaged chemical pesticide containers have reportedly leaked into a nearby dry riverbed that is used for the meager water supply of Hargeisa, capital of Somaliland (Adam, 1995). But before we look abroad, let us examine who benefits and who suffers from corporate decisions in the United States.

Tension Between Culture and Power

The dispute between the Hopi, the Navajo and the Peabody Western Coal Company in Arizona is a perfect case to illustrate how cultural values of indigenous societies are often forgotten in struggles with multinational corporations. The ways in which the Hopi get and use water are a major part of their identity, religious beliefs, ritual practices and daily engagements (Whitely and Masayesva, 1998, p. 12). *Paahu,* "natural water" or "spring", is absolutely central in Hopi social and environmental thought. Every spring is a place of worship and hence a shrine (Fewkes, 1906, p. 346). No spring in the region is without evidence of many offerings to the deities of water. Sacred springs may be regarded as altars and the offerings as sacrifices; whole essence may be carried by the water (Hough, 1906, p. 163). So how does this view differ from a powerful multinational such as Peabody's coal-producing company, whose profit on coal sales was in excess of $ 2 billion (Hanson Industries, 1996)? The Peabody Western Coal Company uses drinking-quality water to transport coal slurry to the Mohave Generating Station in Nevada. They are the only mine in the United States that transports coal by slurry. Was this decision based on economics, environmental racism, or other factors? Nevada and California residents to maintain a lifestyle rich with electronic conveniences such as Jacuzzis, microwaves and heated pools use the power generated at the Mohave Generating Station from this water. Should one culture be sacrificed to feed another culture that is based on consumerism and wealth? The legal mechanism available to the Hopi appears to have failed them in spite of the fact that there were petitions signed by several hundred Hopi citizens at public hearings, dissenting interpretations by independent geologists, an Environmental Objections-Insufficient information classification by EPA, and repeated refusals by the Hopi Tribal Council to sanction Peabody's mining lease (*Proposed permit application, Black Mesa-Kayenta Mine,* 1990).

Are the rules of the game the same if the players have different powers? Are all players equally represented? Many Americans mistakenly place a false sense of security with the Environmental Protection Agency, an agency grossly understaffed, under funded, and, until the passage of the Federal Facilities Compliance Act, with no regulatory control over Department of Defense or Department of Energy sites. In the face of scarce public resources, public expenditures cannot address all environmental and health issues (Bowen, 2001).

Also, the EPA was not designed to address environmental policies that result in unfair practices (Bullard, 1994). In the Hopi case, the EPA's review comments on the mine permit did not have enough impact to have the permit denied. The EPA's comment of the N-aquifer modeling was:

> EPA believes that the available data do not support statements in the Draft Environmental Impact Statement that the cumulative effects of current and foreseeable mining and related operations (principally the coal transport slurry) are expected to result in only minor hydrological impacts (*Proposed permit application, Black Mesa-Kayenta Mine*, 1990, p. 267).

An ironic alternative to the slurrying of aquifer water has been proposed: the construction of another pipeline from Lake Powell that would provide water for domestic use by the Hopi and industrial use by Peabody. We begin by diverting water from one source, thus altering a simple way of life, transporting it across great distances only to return it again. The question remains, will the Hopi celebrate by placing prayer sticks near faucets, will traditional pilgrimages to springs merely follow the pipelines to motorized houseboats docked at Lake Powell?

This case brings to mind the inadequacy of the regulatory system in regard to its tendency to favor big business. Regulating agencies also tend to have a bias for "objective scientific" testimony in processes that are meant to give voice to concerned citizens. To further complicate matters, most citizens do not know about provisions in major laws that allow for public comment. The EPA does maintain a Website, but without legal training a citizen begins with the disadvantage of not knowing how to access available information about the siting of landfills, the status of permit applications, or the existence of environmental impact statements that document major federal regulations affecting the environment. Table 1 presents a list of applicable citations of U.S. environmental statutes and regulations.

Table 1: Research Citations

Environmental Concerns	Laws	General Citation and (Public Comment Citation)	Resource Hotline Numbers Websites
Air discharges	Clean Air Act	40 CFR 50-80	www.epa/gov (databases)
Water discharges	Clean Water Act Safe Drinking Water Act	40 CFR 100-140 400-470 40 CFR 140-149	1-800-426-4791 1-202-382-5533 www.epa/gov
Industrial sites and abandoned waste sites	Resource Conservation and Recovery Act; Comprehensive Environmental Response, Compensation and Liability Act	40 CFR 240-280 40 CFR 300	1-800-424-9346 1-202-382-3000 www.epa.gov (National Priority List)
Industrial facility Permit issues Toxic releases	Resource Conservation and Recovery Act; Emergency Planning and Community	40 CFR 240-280	www.epa.gov (envirofacts/ nviromapper)
		Right to Know Act	
Government projects or Federally funded projects (e.g. logging permit)	National Environmental Protection Act	40 CFR 1500	Environmental Quality Council
Toxic Spills	Emergency Planning and Community Right to Know Act (Lists allowable toxic discharges by facilities in certain neighborhoods)	40 CFR 355	1-800-424-8802 1-202-426-2675

The first step is to know the system; the second step is to re-examine the foundation of our system and other systems around the world. Inequitable issues were brought to the attention of the EPA in 1990 by the Michigan Coalition (a group of social scientists, political activists and biological investigators) (Bryant and Mohai, 1992). The EPA (1992) was shamed into conducting a study that revealed that:

1. There is a clear difference between racial groups in terms of disease and death rates; however, there is a general lack of data on environmental health effects by race and income. The notable exception is lead poisoning. Lead poisoning, while completely preventable, is one of the most common environmental health diseases in the United States. According to the Agency for Toxic Substances and Disease Registry's (ATSDR) report, 49% of African American inner-city children are exposed to dangerous levels of lead (which results in low attention spans, limited vocabulary, and behavior problems) compared to 16% of white inner-city children (*Nature and Extent of Lead Poisoning in Children in the United States*, 1988).
2. People of color and low-income populations experience higher than average exposure to air pollutants, hazardous waste facilities, contaminated fish and farm pesticides in the workplace.
3. Data are not routinely collected on health risks posed by multiple industrial facilities, cumulative and synergistic effects, or multiple pathways of exposure.
4. American Indians are a unique ethnic group with a special relationship to the federal government and have distinct environmental problems. Tribes generally lack physical infrastructure, institutions, trained personnel, and resources necessary to protect their members.

President Clinton issued the Executive Order on Environmental Justice in 1994. This executive order focuses federal attention on the environmental and human health conditions in people of color and low-income populations with the goal of achieving equal environmental protection for all communities regardless of their race, income status, ethnicity, or culture. However, executive orders are mainly directives for government agencies, not the industrial facilities that are the culprits in many environmental justice disputes. No major legislation has been passed, with the exception of state laws in Arkansas and Louisiana.

The EPA does provide financial assistance to communities involved in environmental justice cases. However, many of the organizations do not know how to access the funds. Information on this assistant can be obtained by contacting millard.margaret@epa.gov or www.epa.gov. The EPA has been criticized on their environmental justice record, however, they have limited authority. They are a regulatory agency, and as such, they administer laws set by Congress. People, through voting or issuing public comments, can impact the laws and control the siting of industrial plants and hazardous waste sites. If a new plant is being built in a community or a hazardous waste landfill is proposed, the owner or operator must apply for a Resource Conservation and Recovery Act Part B permit. This permit requires public notification and solicits community input. Concerned citizens can contact the EPA at the telephone numbers listed in Table 1 for more information. Furthermore, all toxic discharges from operating plants must be reported and this information can be accessed through EPA's Website. An individual can find information about discharges or abandoned hazardous waste sites by simply typing in their zip code on the EPA's EnviroMapper (http://maps.epa.gov/enviromapper/). The EPA also employs a hazardous waste ombudsman in Washington, D.C., to respond to citizen grievances.

Race or Business as Usual

Racial and ethic communities suffer the most severe environmental pollution (Bullard and Wright, 1987, p. 12). For example, air pollution levels in the Washington, D.C., metropolitan area were found to be higher in the poorer areas of the city and where the African American population lives. A similar situation exists in New York, Chicago, Denver, Los Angeles and San Francisco (McCaull, 1976, p. 26; Gelobter, 1988). The question arises: Is it a case of racism or economic incentives? Corporations know that, on average, it is about six times more expensive to violate waste laws in white communities than in communities of color (Taylor, 1997, p. 48). If a company can operate in an area where fines, pollution controls, and penalties are based on existing conditions and property values, why would they not, from a purely economic view, locate in a poor neighborhood? Plus, the labor force is willing to accept higher risks and lower pay.

Sexism and Environmentalism

Discrimination is compounded when we consider economics, power and gender issues. Similar discrimination practices can be seen in nearly every country in the world. Women across nations have devoted countless hours to the cause, formally through conferences like *Beyond Beijing* to informal meetings in small villages. Women make up half the world's population but less than 5% of the world's heads of state, chief executives of major corporations, and top positions in international organizations (*Women, Equity & Sustainable Development*, 1998). Yet women play a predominant role in environmental actions. As Vandana Shiva (1992) explains:

> Women seem to realize that there is another world from which different values come, which is the ecological world. Women are able to maintain contact with that world even after the society, the economy, science and technology have tried to make us think that the only economy that matters is the one in which we earn money and the only science and technology that matters is the one we have built (p. 84).

Over half of the environmental groups of color in 22 out of 39 jurisdictions in the United States had women leaders (Taylor, 1997, p. 58). Notable women that have been instrumental in the environmental justice movement include Dollie Burwell from Warren County (toxic waste dump), Linda Campell (who organized the Alabama environmental coalition against the largest U.S. toxic waste dump), and Shelela Cannon (Concerned Citizens of South Central Los Angeles). Scores of women were involved in the Kenya Environment and Energy Organization that promoted a program to use wood and charcoal stoves that reduce fuel. Authors such as Marianne LaVelle and Marcia Coyles (*Unequal Protection* and *The Racial Divide in Environmental Law*) provide an extensive analysis of the discriminatory impacts of environmental policies. For example, communities like the predominately African American Atgeld Gardens (a.k.a. "toxic donut") have 50 toxic sites within a six square mile area, yet the Hazardous Ranking System (HRS) does not take into account the combined effects of these sites on the 10,000 residents (Taylor, 1997, p. 47). Given that the EPA is underfunded and understaffed, rather than criticize the system, a solution could be to apply for an environmental justice grant or privately funded grant with community members, environmental academics, college environmental science or technology consultants, and computer models to revise the HRS, which is used to determine eligibility for superfund clean-up money. Communities have to use their resources, such as educational institutions, to assist the regulatory agencies. Service learning programs could be institutionalized so that students and faculty can receive credit for their environmental stewardship work. For example, in January 1995 the University of Pennsylvania offered a seminar to address the pervasive problem of the exposure of urban children to environmental lead (Giegengack et al., 1999). This project received funding from the EPA and the Kellogg Foundation.

Ecofeminism

Let us revisit the connections between women and environmental destruction. Environmental degradation affects women disproportionately because women's activities include the acquisition of water and fuel wood. As deforestation increases, women are forced to travel greater distances. In South Africa, women generally spend seven to nine hours per week on these chores, walking 12 to 38 miles per headload. Headloads as heavy 147 pounds have been recorded. This means less time for childcare, education, paid work, or other activities. In a country where women are not entitled to legal protection until menopause, it is questionable if their concerns are heard (Goodland, 1995).

Women's knowledge is ignored at great costs. When Chipko villagers in the Himalayans were asked about which trees should be planted as part of a reforestation plan the men immediately choose fruit trees. The women responded, "The men would take the fruits and sell them by the roadside. The cash would go to buy liquor and tobacco. We women prefer fuel and fodder trees" (Dankelman and Davidson, 1988).

Women cannot afford the illusion that it is possible to escape time and place. This is one reason for calling typically women's knowledge "expert knowledge" (Curtin, 1992). Women are linked more closely temporally and spatially to their surroundings. The link between women and the environment stems from experiencing the effects of domination is it of nature or of themselves (Merchant, 1980; Breton, 1998, p. 213; Hamilton, 1990; Adair, 1990).

Age and Environmental Discrimination

Another form of environmental discrimination not generally addressed is age discrimination. In many older American inner-cities, industries locate in mixed zoning areas because older residents are less likely to protest. Older residents are less informed about how to address the health issues although they are a population at risk. Often they feel a strong connection to their neighborhoods because of many years that they have lived there.

Simple Steps towards Justice

Public education is a critical step in addressing environmental justice concerns. For example, these are simple solutions to the lead contamination problem:

- Use phosphates dishwater powder that has proved useful in cleaning up lead dust.
- Exposed individuals should supplement their diets with iron and calcium.
- Pregnant women and children should not enter renovated homes for 24 hours.
- Apply to free testing programs through local Department of Health or Medicaid offices.
- Volunteer to repaint older low-income homes for the elderly or single mother households with small children.

Personal responsibility outside and within the home is important because certain actions have long-term consequences. For example, the period of time between an exposure and the development of chronic effects can range from 10 to 20 years. This becomes a problem in terms of environmental litigation. Twenty year olds often accept work not realizing that years later, when they want to start a family, they will not be able to. Although the analytic capability and technology is available to detect the health effects of minute concentrations of many chemicals, only 7,000 of the 5 million substances known have been tested for carcinogenicity or other deleterious effects, such as decreased fertility (Bowen, 2001). Rather than using less toxic alternatives, Americans often use household chemicals and pesticides on their lawns that can have adverse effects on health. The effects of these choices add up. In the case of many hazardous chemicals, the availability of data with which to measure the component of risk is rare (Bowen, 2001). Also, some community members (i.e. children) are not cognizant of risks. The lifelong impact of early childhood exposure to lead is difficult to quantify but its contemplation has led a generation of public health practioners to agree that low-level lead poisoning is the single greatest environmental hazard that urban children face today and that hazard disproportionately threatens African American children (Giegengack et al., 1999).

We need to recognize how our moves affect others and try to understand our common concerns. We live in a country that represents 6% of the world's population and consumes 45% of its resources, including 60% of its energy resources (Hamilton, 1990, p. 63). According to environmental philosopher Eugene Hargrove (1995), one reason that environmental racism has been ignored is that that environmental ethics literature, for the most part, reflects environmentalists' concerns, which have not included concerns about human welfare. The environmental justice movement is meant to unite environmentalism and social justice by challenging the business-as-usual environmentalism that is generally practiced by the more privileged wildlife and conservation-oriented groups (Bullard, 1993). For years, the environmental movement operated without significant minority participation (Buttel and Flinn, 1974; Loew, Pinhey and Grimes, 1980; Fox, 1985). Some white environmentalists dismiss this new sector of the environmental movement as radical extremists, while others have sought to make just enough changes within their organizations to avoid charges of racism. Both groups need to identify the issues that keep them apart as well as the radical changes needed to bridge the distance (Bullard, 1993).

All environmentalists should strive to find common ground and build alliances that can gain political support. It is important to remember that people are concerned with their immediate surroundings or where their interests lie; a place where they feel a connection—perhaps a park they walk through or bring their children to watch birds. A quiet place where they reflect on their lives and remember:

> The world can tell us everything we want to know. The only problem for the world is that it doesn't have a voice. But the world's indicators are there. They are always talking to us— Iwjivik Quitsak Tarkiasuk (cited in Poslums, 1993).

In conclusion, it is critical that more people utilize the environmental legal and scientific systems and, more importantly, exert their voice to impact environmental decision-making regardless of the personal, academic, or professional implications. We cannot change the past but we can change the future by seeing humans and the non-human world justly and by living simply. All humanity can have an impact through the process of setting an ethical environmental example in some capacity.

References

Adair, M. (1990). "Embracing Diversity: Building Multicultural Alliances", in B. Erickson (ed.), *Call to Action: Handbook for Ecology, Peace and Justice* (pp. 42-46). San Francisco, CA: Sierra Club Books.

Adam, H. (1995). "Somalia: Environmental Degradation and Environmental Racism", in L. Westra and P.S. Wenz (eds.), *Faces of Environmental Racism: Confronting Issues of Global Justice* (pp. 181-207). Lanham, MD: Rowan and Littlefield.

Bowen, W. (2001). *Environmental Justice Through Research-Based Decision-Making*. New York, NY: Garland Publishing.

Breton, M.J. (1998). *Women Pioneers for the Environment*. Boston, MA: Northeastern University Press.

Bryant, B.I. and Mohai, P. (1992). *Race and the Incidence of Environmental Hazards: A Time for Discourse*. Boulder, CO: Westview Press.

Bullard, R.D. (1990). *Dumping in Dixie: Race, Class, and Environmental Quality*. Boulder, CO: Westview Press.

Bullard, R.D. (ed.). (1993). *Confronting Environmental Racism: Voices from the Grassroots*. Boston, MA: South End Press.

Bullard, R.D. (1994). "Overcoming Racism in Environmental Decision Making", *Environment,* 30(4): 11-20. 39-44.

Bullard, R.D. and Wright, B.H. (1987). "Environmentalism and the Politics of Equity: Emergent Trends in the Black Community", *Mid-American Review of Sociology,* 12, 21-32.

Buttel, F. and Flinn, W. (1974). "The Structure and Support for the Environmental Movement 1968-1970", *Rural Sociology* 39, 56-69.

Carson, R. (1962). *Silent Spring*. Boston: Houghton Mifflin.

Chavis, B. (1993). "Toxic Waste and Race", in R.D. Bullard (ed.), *Confronting Environmental Racism: Voices from the Grassroots* (pp. 3-5). Boston, MA: South End Press.

Curtin, D. (1992). *Women's Knowledge, Expert Knowledge*. Paper presented at a meeting of the International Development Ethics Association, Honduras.

Dankelman, I. and Davidson, J. (1988). *Women and Environment in the Third World: Alliance for the Future*. London: Earthscan.

Environmental Defense Fund (2003 January 1). *Senate Roll Call Votes for Amendments Relating to the Weakening of the Clean Air Act's Pollution Rules*. Retrieved from http://www.environmentaldefense.org/pdf.cfm? Content ID=2626 & File Name=Senate NSR Votes Jan03 % 2E pdf.

Environmental Protection Agency (1992). *Environmental Equity: Reducing the Risk for all Communities* (EPA230-R92-008). Washington, DC: U.S. Environmental Protection Agency.

Environmental Protection Agency (1994). *Environmental Justice Initiatives 1993* (EPA 200-R-93-001). Washington, DC: U.S. Environmental Protection Agency.

Environmental Protection Agency (2003). "Federal Insecticide, Fungicide and Rodenticide Act, as Amended", *40 US Code of Federal of Regulation, Part 162*.

Fewkes, J. (1906). "Hopi Shrines Near the East Mesa, Arizona", *American Anthropologist,* 8, 346-375.

Firestone, D.B. and Reed, F.C. (1993). *Environmental Law for Non-Lawyers*. South Royalton, VT: SoRo Press.

Fox, S.R. (1985). *The American Conservation Movement: John Muir and His Legacy*. Madison, WI: University of Wisconsin Press.

Gelobter, M. (1988). *The Distribution of Air Pollution by Income and Race*. Paper presented at the Second Symposium of Social Science in Resource Management, Urbana, IL.

Gibbs, L.M. (1982). *Love Canal: My Story*. Albany, NY: State University of New York Press.

Giegengack, R., Cressler, W. and Bloch, P. (1999). "An Educational Strategy to Reduce Exposure of Urban Children to Environmental Lead: ENVS 404 at the University of Pennsylvania", in H. Ward (ed.), *Acting Locally: Concepts and Models for Service-Learning in Environmental Studies* (pp. 19-26). Washington, DC: American Association for Higher Education.

Goodland, R.J.A. (1995). "South Africa: Environmental Sustainability Needs Empowerment of Women", in L. Westra and P.S. Wenz (eds.), *Faces of Environmental Racism: Confronting Issues of Global Justice* (pp. 207-227). Lanham, MD: Rowan and Littlefield.

Hamilton, C. (1990 April 1). "Women, Home, Community", *Race, Poverty & the Environment Newsletter*.

Hanson Industries (1996). *Annual Report for 1996*. Woodbridge, NJ: Hanson Industries.

Hardin, G. (1968). *The Tragedy of the Commons*. Washington, DC: American Association for the Advancement of Science.

Hargrove, E. (1995). "Foreword", in L. Westra and P.S. Wenz (eds.), *Faces of Environmental Racism: Confronting Issues of Global Justice* (pp. ix-xiii). Lanham, MD: Rowan and Littlefield.

Hough, W. (1906). "Sacred Springs of the Southwest", *Records of the Past*, 5(6): 163-169.

Houlihan, J., Campbell, C. and Wiles, R. (2000). *Moms... and POPs: Persistent Organic Pollutants in the Diets of Pregnant and Nursing Women*. Washington, DC: Environmental Working Group. Retrieved from http://www.ewg.org/reports/momsandpops/pops.html

Jenks-Jay, N. (1999). "Connecting the Classroom and the Community: A Southern California Experience", in H. Ward (ed.), *Acting Locally: Concepts and Models for Service-Learning in Environmental Studies* (pp. 41-49). Washington, DC: American Association for Higher Education.

Kaufman, D.G. and Franz, C.M. (2000). *Biosphere 2000: Protecting Our Global Environment*. Dubuque, IA: Kendall/Hunt.

Knight, D. (2000). "Displacing Kurds, Drowning the Past", *World Press Review* 47(12): 21-22.

Kokole, O. (1995). "The Political Economy of the African Environment", in L. Westra and P.S. Wenz (eds.), *Faces of Environmental Racism: Confronting Issues of Global Justice* (pp. 163-181). Lanham, MD: Rowan and Littlefield.

Lowe, G., Pinhey, T. and Grimes, M. (1980). "Public Support for Environmental Protection: New Evidence from National Surveys", *Pacific Sociological Review,* 23, 423-445.

McCaull, J. (1976). "Discriminatory Air Pollution: If Poor, Don't Breathe," *Environment,* 18(2): 26-31.

Merchant, C. (1980). *The Death of Nature: Women, Ecology, and the Scientific Revolution.* San Francisco, CA: Harper and Row.

The Nature and Extent of Lead Poisoning in Children in the United States: A Report to Congress (1988). Atlanta, GA: Agency for Toxic Substances and Disease Registry, Public Health Service, U.S. Dept. of Health and Human Services.

Pardo, M. (1990). "Mexican American Women Grassroots Community Activist: Mothers of East Los Angeles", *Frontiers: A Journal of Women's Studies,* 11, 1-6.

Phillips, R. (1995). "Evanston Community and Environmental Racism: A Case Study in Social Philosophy", in L. Westra and P.S. Wenz (eds.), *Faces of Environmental Racism: Confronting Issues of Global Justice* (pp. 93-113). Lanham, MD: Rowan and Littlefield.

Poslums, M. (1993). *Voices from the Odeyak.* Toronto: NC Press.

Proposed Permit Application, Black Mesa-Kayenta Mine, Navajo and Hopi Indian Reservations, Arizona: Final Environmental Impact Statement. (1990). Denver, CO: U.S. Office of Surface Mining Reclamation and Enforcement.

Shiva, V. (1984 May 19). "Acute Water Crisis Grips Uttar Pradesh", *Indian Express.*

Shiva, V. (1989). *Staying Alive: Women Ecology and Development* (2nd ed.). London: Zed Books.

Shiva, V. (1992). "Interview", in S. Lerner (ed.), *Beyond the Earth Summit: Conversations with Advocates of Sustainable Development.* Bolinas, CA: Common Knowledge Press.

Tarlock, D. (1994). "City versus Countryside: Environmental Equity in Context", *Fordham Urban Law Journal,* 21, 461-494.

Taylor, D.E. (1997). "Women of Color, Environmental Justice and Ecofeminism", in K. Warren and N. Erkal (eds.), *Ecofeminism: Women, Culture, Nature* (pp. 38-81). Bloomington, IN: Indiana University Press.

United Church of Christ Commission for Racial Justice (1987). *Toxic Wastes and Race in the United States: A National Report on the Racial and Socio-Economic Characteristics of Communities with Hazardous Waste Sites*. New York: United Church of Christ.

United Church of Christ Commission for Racial Justice (1992). *The First National People of Color Environmental Leadership Summit: Program Guide*. New York: United Church of Christ.

Westra, L. (1993). "A Transgenic Dinner? Ethical and Social Issues in Biotechnology and Agriculture", *The Journal of Social Philosophy*, 24(3): 215-32.

Westra L. and Wenz, P.S. (eds.). (1995). *Faces of Environmental Racism: Confronting Issues of Global Justice*. Lanham, MD: Rowan and Littlefield.

Whitely, P. and Masayesva, V. (1998). "The Use and Abuse of Aquifers: Can the Hopi Indians Survive Multinational Mining?" in J.M. Donahue and B.R. Johnston (eds.), *Water, Culture and Power: Local Struggles in a Global Context* (pp. 9-35). Washington, DC: Island Press.

Wipke, K.B. (1998). *Improving Fuel Economy of Sport-Utility Vehicles (SUVs) Through Diesel Technology and Vehicle Improvements.* Center for Transportation Technologies and Systems. Retrieved from http://www.ctts.nrel.gov/analysis/pdfs/SUV.pdf.

Women, Equity & Sustainable Development (Teacher's Guide to World Resources). (1998). Washington DC: World Resource Institute.

World Commission on Environment and Development (1987). *Our Common Future*. New York, NY: Oxford University Press.

The Modern History of Energy Conservation: An Overview for Information Professionals

12

Donald R. Wulfinghoff

Energy conservation is one of the critical issues facing society today. Our civilization runs on energy. However, energy resources are finite. Increasing demand is being made for diminishing supplies. The cost of energy is enormous, and the cost is rising. Utility bills account for much of the cost of housing, and they are a major cost of business. A large hotel or hospital spends millions of dollars for energy each year. Some steel mills pay hundreds of millions of dollars annually for energy. This consumption brings a host of environmental dangers. Fossil fuels dump carbon dioxide into the atmosphere, accelerating the greenhouse effect. Air conditioning systems release gases that destroy the earth's ozone layer. Discarded lamps contribute to mercury pollution. Energy consumption depletes not only the fuels themselves, but also a spectrum of other resources. A fundamental issue today is providing information about energy conservation. It is a special challenge for information professionals because:

- Energy conservation relates to many areas of human activity, so the original information about efficiency is scattered throughout the literature of many fields.
- By the same token, information about energy conservation is often subsidiary to other topics.
- Energy conservation is still evolving as a distinct field of knowledge. Therefore, even modern information about it, is scattered.

- The best information often comes from sources that are unconventional for librarians.
- Information about energy conservation varies widely in quality and credibility.
- Attempts to create directories of information about energy conservation have failed to keep up with the quantity and variety of information.
- The information client often has no clear idea of what he wants to learn.

There is a vast sea of information about energy conservation. From the standpoint of the information professional, this sea is still largely uncharted and full of strange hazards. The purpose of this article is to help everyone to navigate this sea and find needed information. The present organization of information about energy conservation has been shaped by the way that energy conservation has evolved. Therefore, the author discusses the historical currents of information about energy conservation and stresses the modern era, which began in 1973. Then he views the overall geography of energy conservation information, followed by field trips to the most important groups of information. Finally, to avoid giving misinformation to information clients, the author examines some of the strange customs and beliefs that exist in the lands of energy conservation. At the end of this guide he includes a selective list of Internet sites.

What is Energy Conservation?

"Energy conservation" can mean a variety of things, and the most common meanings are:

- Using less energy in a particular application.
- Finding ways to purchase particular forms of energy at lower cost. This is usually accomplished by negotiating with energy providers or by using energy under less costly conditions. (Paradoxically, the latter method may increase energy consumption considerably.)
- Shifting to different energy sources of lower price.

- Using "free" or "renewable" energy sources (Paradoxically, this is often expensive).
- Shifting to energy sources that are considered to be more desirable, or less undesirable, with regard to non-efficiency concerns such as availability and pollution. Such shifts typically involve serious compromises.
- Conserving water and materials, as well as energy sources.

Information about energy conservation is not usually grouped in ways that distinguish between these meanings. In this article the author will treat all of them under the mantle of "energy conservation", "energy efficiency", and "energy management", and use similar terms interchangeably. These terms have no standardized meanings at the present time. Energy conservation is now widely recognized as a fundamental strategy for protecting the environment. Therefore, expect to find more and more information about energy conservation under the heading of "environmental protection" and related terms.

The Early History of Energy Conservation

The imperative to conserve energy is as old as the use of energy. For most of human history, use of energy was limited to the amount of work that could be done by human beings, usually alone, but sometimes in large groups. Later, humans learned to use animals and teams of animals to do the tasks requiring heavy lifting and hauling. Neither humans nor animals like hard work. Aversion to work strongly motivated energy conservation from the beginning. Energy conservation first consisted of doing less. Then, as intelligence evolved, it included finding easier ways to get work done. For example, the invention of the wheel was an early advance in energy conservation. Fire is the oldest major source of energy, other than muscle, that is controlled by humans. Since prehistory, fire has been used for cooking, heating of dwellings, hardening spear points, clearing land, smelting and casting metal, baking pottery and other applications. Controlled fires require a considerable amount of effort for gathering fuel, so efficiency arose in the use of fuel. For example, the kilns and ovens of many early cultures were quite efficient in their use of fuel. Also, North American Indians devised ways of using far less fuel for comfort heating than European immigrants.

Wind power is the oldest major source of mechanical energy, other than muscle. Its earliest major application was ship propulsion, which began perhaps five to ten thousand years ago. This made possible the trading of large tonnages of goods throughout the known world. Moreover, wind power expanded the boundaries of the human world and it was an essential factor in the development of many civilizations.

Land-based rotary wind machines have been used in a limited number of cultures for over two millennia. Efficiency evolved by trial and error. The Dutch style of wind machine used for pumping water and grinding grain five hundred years ago achieved a large fraction of the theoretical maximum efficiency. However, usage is limited by the localized and irregular nature of wind.

The energy of falling or flowing water has been used to a significant extent for over two millennia, becoming a major energy source in the Middle Ages. Unlike wind, which is a nuisance in itself, water is desirable for many important purposes. As a result, many cultures settled near water supplies, which they learned to exploit for power. Well designed "overshot" water wheels built 500 years ago have efficiencies that approach the theoretical maximum. Water power is limited by the amount of stream flow and by the height of the fall that is readily available. The technology of building dams tall enough to augment power production was not highly developed until the 19th century.

Steam machinery is another application of fire. It was first put into practical application during the 18th century. Its development was revolutionary for several reasons. It enables fire to produce mechanical work. It was the first invention capable of producing mechanical power in any locale, on land or sea. Virtually any amount of work can be done by building larger machines. Steam engines can operate whenever work is needed, independent of climate. To a large extent, steam power created modern civilization. However, steam machinery requires fuel, unlike wind and water power. The ability of mechanics to build large steam engines, and the growth in applications for such engines, led to rapidly increasing demand for fuel, which could no longer be satisfied by whatever excess wood happened to be in the vicinity. Fuel cost soon emerged as a dominant limitation to the application of steam machinery.

In response, efficiency also emerged early as a fundamental issue in the design of steam machinery. In fact, during the 18th and 19th centuries, what we now call mechanical engineering was largely a search for higher efficiency. James Watt and many other engine builders improved efficiency on an empirical basis, while attempting to understand the underlying physics. Sadi Carnot created the first solid theoretical understanding of energy efficiency. During the early 19th century, Carnot explained the Second Law of Thermodynamics, which places a severe theoretical limit on the efficiency of converting heat to mechanical work.

New fuel sources were discovered and developed in parallel with the development of power producing machinery. This was largely a fortunate coincidence. It was not primarily a matter of new machinery creating a demand for new energy sources. On the contrary, from the beginning of the 19th century, new sources of fuel were developed or discovered in advance of demand. Steam machinery made coal mining much more productive. The extraction of petroleum in quantity was developed in the middle of the 19th century, before people knew what to do with large amounts of petroleum. The development of internal combustion engines soon provided a large market for petroleum, which later found application as a primary fuel for steam machinery. With petroleum came natural gas, initially a dangerous waste product. Its discovery in quantity again motivated people to develop uses for it.

Electrical power first emerged in the late 19th century, specifically for lighting. Electrical power was produced by increasingly efficient engines. However, lamps remained inefficient until the commercialization of fluorescent lighting, shortly before World War II. The development of practical electric motors, largely by Nikola Tesla, occurred toward the end of the 19th century. This enormously expanded applications for mechanical power. It freed the individual energy user from the need to have his own steam plant. Mechanical power could now be used anywhere, even in very small applications. The invention of innumerable small machines and labor saving devices made "energy" a ubiquitous commodity by the beginning of the 20th century.

Unlike the evolution of mechanical equipment, the development of electrical equipment was largely based on theory. All practical electrical motors are efficient, at least in comparison with combustion-driven machinery. However, the efficiency of applications served by inexpensive alternating-current motors is often limited by the fact that these motors are single-speed devices. Efficient variable-speed motors were developed early, but they had serious cost and maintenance limitations.

By the beginning of the 20th century, energy consumption *per capita* was accelerating, while the energy-consuming population of the earth also grew rapidly. Appliances displaced muscle power at home. Machines increased production in factories and in agriculture. Automobiles made transportation a major new consumer of fuels. Fuel replaced wind for the propulsion of ships. Air travel became another user of fuel. All the while, the available supply of energy continued to grow comfortably ahead of demand. Huge hydroelectric generation plants were built to provide jobs during the Depression. Electricity generation by nuclear fission arose as a byproduct of nuclear weapons, becoming another major source of energy from the 1950's onward.

Until the early 1970's, there was a popular conception of continually diminishing energy prices. For example, nuclear power advocates spoke of electricity that would be "too cheap to meter". As a result, efficiency ceased to be a major concern of the engineers who designed energy-using equipment, and efficiency faded as an issue with the public and the government.

The Modern Era of Energy Conservation

The Big Bang that started the modern era of energy conservation was the "energy crisis" that erupted in 1973. This resulted from an embargo of oil that was directed primarily against the United States by Arab oil producing countries in reaction to U.S. policy in the Middle East. Energy prices in the United States rose dramatically, mainly for petroleum. Far more compelling, there were actual shortages of petroleum, mainly in the form of annoying lines at gasoline stations. And, there were threats of insufficient heating oil.

In fact, petroleum prices in the United States never reached the levels that had long been routine in other countries. The big change introduced by the 1973 "energy crisis" was the realization that energy sources might not keep pace with mankind's ability to use energy. This was not a new concept for specialists in energy resources, but it was new as a popular idea. Stated differently, the supply of energy was no longer viewed as something that was always ahead of demand. Instead, the supply of energy, although still vast, was now viewed as lagging demand.

Previously, energy efficiency had been a technical aspect of designing equipment, systems and buildings. In 1973, efficiency metamorphosed into "energy conservation", which emerged as a distinct field of interest, rather than continuing to be a subsidiary engineering issue. Energy conservation became a single freestanding issue, independent of the many technical areas to which efficiency applies specifically. This single-issue nature of energy conservation became a dominant factor in information about energy conservation from that time until the present.

The differences between the new era and the early era of energy conservation are so important that we will summarize them:

Old Energy Conservation	New Energy Conservation
Energy sources are discovered ahead of demand.	Energy sources are being depleted, without replacement.
An issue for individual parties.	A societal issue.
A technical aspect of individual machines and processes.	A freestanding issue, struggling to become technically integrated.
Perceived by everyone as an issue of labor requirements and fuel costs, to be addressed by technical and economic means.	Perceived mainly as a resource conservation and/or environmental issue. Distinct interest groups variously seek to address it by economic, technical, political, social and/or metaphysical means.

The Major Currents of Modern Energy Conservation

Energy conservation is an issue with many aspects that continue to evolve. There have been major areas of technical improvement. There are also important areas in which there has been virtually no improvement, and hence, most of the potential of energy conservation still remains to be tapped. In the architectural arena, there has been serious efficiency regression, primarily related to the use of glass box exterior design.

The United States was the primary target of the oil embargo, and in the United States, popular opinion drives politics. As a result, there were many government initiatives toward energy conservation, including massive spending on research, major new laws, and incentives for favored energy conservation activities. Energy conservation continues to be a national imperative.

These are significant generalizations that we can make about the modern era of energy conservation, thus far:

- The 1970s were a time of great ferment and rapid learning. The 1980s was a period in which many bad ideas from the 1970s collapsed. The 1990s was a time of stagnation and information loss. The new millennium restores interest in energy conservation, largely under the banner of environmental protection.
- By far the greatest advances came from equipment manufacturers. Most new concepts were introduced

during the 1970s. By the 1990s, the equipment had become reliable. Remaining technical development is slow.

- Architects have disregarded efficiency in building design. Engineers, who are subordinate to architects in building design, continue to improve efficiency in a desultory manner.
- Most government programs failed. Important exceptions were information programs and equipment efficiency standards.
- Energy conservation codes for construction were enacted into law, but they remain widely ignored.
- The industrial sector achieved much better efficiency improvement than the commercial sector, achieving most improvement during the years 1974 to 1985. Progress since then has been minor.
- Facility owners and managers do not yet recognize energy efficiency as a normal aspect of management, but as an episodic issue.
- During the first part of the modern era, energy conservation was independent of environmental protection. Then, they were seen as conflicting issues. Now, they are seen as complementary issues.
- Information mushroomed during the 1970s, and slowed substantially thereafter. The *Energy Efficiency Manual* finally appeared at the beginning of the new millennium, filling the need for a complete and easily usable guide to the main areas of energy conservation.

All this development and turmoil has created strong currents of issues within the modern era of energy conservation. These are summarized in Figure 1.

The Geography of Energy Conservation Information

It helps to view information about energy conservation in these groups and sub-groups:

- the primary grouping is by energy end-use sector

- within each end-use sector, there are sub-groups of information that are targeted to the practitioners in those sectors
- also within each end-use sector, there are sub-groups of information that relate to the equipment and materials used in that sector
- a separate body of policy and advocacy information exists largely in isolation from the technical literature.

Understanding this large-scale geography helps to start searching for information by sailing in the proper direction. Below one can examine each of these groups in greater detail.

Information Grouped by Usage Sectors

Energy usage by humanity is divided into three large sectors, each of approximately equal size. One-third of the world's energy is consumed in the buildings where we live, work and congregate. Another third is used in the factories and plants that produce our needs. The remaining third is used in transportation.

The term "energy conservation", as it is generally used, applies to the first two end-use sectors, buildings and industry, but not to transportation. (Energy conservation in transportation is considered an integral part of the engineering of the vehicles, or it is a social discussion about different modes of transportation.) The buildings sector is divided into residential and commercial buildings. (The latter includes all large buildings that are intended primarily for human occupancy, regardless of whether the activity is "commercial" or not.) Residential buildings account for about 80% of energy consumption in the buildings sector, primarily because of their great number.

Energy conservation in the industrial sector falls into two distinct groups: conservation methods that are common to most industrial facilities and efficiency improvements to specialized industrial processes. One is likely to find information about the common processes under the heading of "energy conservation". Only a fraction of industries, such as paper making and primary metals, have specialized energy processes. However, some of these are very energy intensive. One will find most information about conservation in the specialized processes within the literature of the particular industry.

In summary, there are four archipelagos of information about energy conservation based on the sectors in which the energy is used:

- residential buildings
- commercial buildings
- the common uses of energy in industry
- improvement of specialized industrial processes.

These archipelagos have virtually no professional or technical connection with each other. There is little commonality in the information sources for them. However, they may share some islands of information about particular topics, such as lighting.

Note that this classification does not include energy that is consumed by the energy or utilities industries. These comprise an intermediate step on the way to the end user of energy. A large amount of energy is lost in energy industries, primarily in the production of electricity by burning fuels. (The laws of physics impose a severe and unavoidable efficiency penalty on this conversion process.) Energy industries tend to be operated as efficiently as possible. Energy conservation is considered an integral part of design, as in the transportation sector. Energy conservation in certain common components of the energy industry, such as cooling towers and lighting, falls within the category of common industrial energy uses.

Information Oriented to Practitioner Groups

Within each of the four end-use archipelagos, there are islands of information that are intended for the different practitioner groups who work in each end-use sector. The information for each practitioner group is produced primarily by the practitioners themselves and by the manufacturers of the equipment with which the practitioners work. Academic sources may produce the educational or theoretical components of practitioners' literature.

The main practitioner groups who influence energy efficiency are:

- facility owners, including homeowners
- facility operating staffs
- facility administrative managers
- architects
- HVAC design engineers (called "mechanical engineers" in commercial sector)
- electrical power engineers
- lighting designers (often a sub-group of architects or electrical engineers)
- industrial process specialists
- builders and contractors
- energy conservation specialists.

Practitioner groups with similar titles may have different bodies of knowledge in different end-use sectors. For example, "mechanical engineers" in the commercial sector are involved exclusively with heating, ventilating, and air conditioning ("HVAC"), whereas "mechanical engineers" in industry may be involved with machine tools, conveyor belts, air compressors, and a wide variety of other matters.

Practitioners in different groups typically do not communicate with each other effectively. As a result, the information for each group may be ineffective or even incomprehensible to the other groups. Therefore, when your client appears with an information request, attempt to find out whether your client belongs to a particular practitioner group, or wants the viewpoint of a particular group. An important exception to the segmentation of literature by practitioner group is the *Energy Efficiency Manual*, which was designed to be comprehensible to readers of all backgrounds. This publication is discussed below.

Information about Equipment and Materials

The energy consumption of a facility is determined largely by the equipment and materials used in it. Fortunately, there are highly developed islands of information related to virtually all types of equipment and materials. In particular, the energy efficiency characteristics of equipment and materials for all end-use sectors are usually well documented. Manufacturers are the primary producers of information about equipment and materials. The information appears first in their catalogs. We discuss this important source below.

There is a great deal of overlap between information sources that are related to professional groups (discussed previously) and information sources that are related to materials and equipment. This is because manufacturers produce their information primarily to induce practitioners to purchase or recommend their products. Professional and academic information sources are also important. Professional literature is more likely to address application problems that manufacturers have not yet solved. Academic literature is closer to the truth about efficiency performance.

Equipment efficiency ratings are another important information source. Efficiency ratings are easy to understand, if you spend a few moments learning how the rating factors are defined. Federal law requires efficiency ratings to be published for major residential energy-using equipment, including heating, cooling and water heating equipment. Certain states, especially California, have efficiency standards for this equipment that are more stringent than the federal standards. Federal law also requires efficiency ratings for certain common types of equipment that are used primarily in the commercial and industrial sectors. Examples are efficiency ratings for particular types of fluorescent lamp ballasts and for the most common types and sizes of electric motors. The Energy Star Program of the U.S. Environmental Protection Agency labels a wide category of products for efficiency.

Trade associations and others may compile equipment efficiency information. For example, the Air Conditioning and Refrigeration Institute compiles efficiency ratings for all air conditioners and heat pumps below a certain size that are made in the United States. The list of websites at the end of this article includes a large number of sites that provide equipment efficiency ratings.

Some government agencies perform an important function in disseminating technical information that has been prepared by others, including many technical studies funded by the government. For example, the U.S. Department of Energy Office of Industrial Technology (OIT) has information programs devoted to specific industries. The OIT also produces a comprehensive computerized tabulation of electric motor efficiency that is called Motor Master.

Information Related to Energy Policy

The previous sources are located in the ocean of technical information. There is another archipelago of information that lies in the ocean of energy policy. The people who inhabit the two oceans come from different backgrounds, and they have different viewpoints and agendas. They communicate with each other poorly, if at all, and there is tension between the two groups. Many people within each archipelago are virtually unaware of the existence of the people and information in the other. There are several important groups of people who deal primarily with the policy aspects of energy conservation, rather than the technical issues. These include government, lobbyists and advocacy organizations.

In the United States, the federal government and some state governments exercise initiative in energy conservation. Legislation is the primary form of government policy information. Legislation arises in ways that are generally not well documented before the legislation appears. Industry newsletters and the popular media may be the best source of information about pending legislation. Although laws are developed as a policy matter, they commonly produce mandates that are technical in nature. For example, an important group of laws create energy conservation codes for building construction and efficiency standards for equipment. The U.S. federal government also makes laws that dictate the selection of fuels in many applications, largely in response to environmental concerns.

Lobbyists typically represent particular industries that may be affected by energy conservation legislation. These parties are generally not a primary source of information about conservation, although they may compile information in a manner that favors their cause. For example, energy industries are a source of such information.

A growing number of advocacy organizations have become involved with energy conservation. A few of these organizations, such as the Alliance to Save Energy and the American Council for an Energy Efficient Economy, have energy conservation as their primary issue. In recent years, the major environmental advocacy groups, such as the Natural Resources Defense Council and the Sierra Club, have embraced conservation as a primary means of environmental protection. Energy conservation may become entangled with other issues, including political, social, or commercial agendas, or some combination of these.

Information that is produced by advocacy organizations is usually derivative. In fact, some organizations make an important contribution by compiling and analyzing existing information. Their output is commonly in the form of articles, conference proceedings and "white papers". Expect this class of information to be heavily flavored to advance the various agendas of the organization, not all of which may be apparent.

On the positive side, compared to other sources, information from advocacy sources tends to be more forward-looking, and it encompasses a broader range of issues. For example, it tends to be the first to raise alarms about environmental dangers. On the negative side, advocacy sources tend to be technically naïve and oblivious to practicality. A common example is promoting the installation of wind generators without regard to the availability of wind. Advocacy sources may display little sense of balance, either between conservation and other issues, or between different methods of conservation (see more about this at the end of this article).

The Main Information Sources

Comprehensive Energy Conservation References

Each well developed field of endeavor has a comprehensive reference that functions as a repository of the information of the field, and it often serves as the textbook of the field. In fact, the appearance of such a reference is often the first step in the emergence of a field as an area of human competence. For example, the advance of medical practice from quackery to an effective science had the publication of major books as its milestones. Similarly, the consolidation of each branch of engineering has been marked by the emergence of a primary reference book in that field.

During the 1970's "energy crisis" years, there were several important efforts to produce comprehensive energy conservation manuals. Hindsight reveals these efforts to have been fragmentary, but they were major advances in making energy conservation a rational subject area. Most of these earlier efforts are classified under the Library of Congress classification TJ163. Unfortunately, there is no corresponding Dewey Decimal Classification number for energy conservation as a whole.

The first truly comprehensive guide to energy conservation arrived at the beginning of the new millennium. It is the *Energy Efficiency Manual* by Donald R. Wulfinghoff, published by Energy Institute Press. It was written explicitly to serve as a comprehensive store of knowledge, and to be user-friendly for anyone involved with energy conservation. To achieve both breadth and depth, the *Energy Efficiency Manual* is big. It has 1,536 pages and 830 illustrations. It presents two types of information. The first 1,200 pages present 400 energy conservation measures, with a variety of aids to make the information quickly and easily accessible. These are written in conversational style, at a "Consumer Reports" level of detail. The last part of the book explains the most important general topics of energy conservation, ranging from brief explanations of terms to in-depth coverage of emerging technologies.

The *Energy Efficiency Manual* does not duplicate extensive engineering design information. For activities where this is needed, the book dovetails with the major professional references. For example, it tells the reader how to improve the efficiency of air conditioning, but it refers the reader to the *ASHRAE Handbook* for design data. The *Energy Efficiency Manual* covers three of the four energy end-use sectors described previously—residential, commercial and common industrial uses. It does not cover specialized industrial processes. For example, in a paper mill, it covers the motors, pumps, fans and lighting, but it does not cover the specialized paper making equipment. For the specialized processes, one needs to be refered to the literature of the particular industry.

Books on Individual Energy Conservation Topics

Books are available about nearly every aspect of energy conservation, except for the most recent developments. Their quality ranges from excellent to fair. Books are the best source of aggregated information, for the reasons that we discuss below. A main challenge will be finding the best books for the needs. We discuss general methods of searching for books below.

There are several publishers that specialize in books about energy conservation, and these are good places to start. The Association of Energy Engineers is the most prolific of these. It is wise to keep a copy of their catalog on hand. There are several vendors that specialize in selling conservation books produced by a variety of publishers. For example, Iris Communications has a fairly good selection of books about residential construction, a large fraction of which involve naturalistic construction methods and renewable resources. The Internet sites of these and other sources are listed at the end of this article. These collective sources account for only a fraction of the energy conservation books that one may wish to find. We will explain below how to search for the energy conservation books individually.

Remember that energy conservation is not clearly distinct from the equipment and processes in which it is applied. This increases the range of sources that you should consider. For example, the famous book *Steam* is not an energy conservation book, but it is a fundamental source for learning about the efficiency aspects of boiler design.

Energy Conservation in the Major Professional References

Despite the emergence of energy conservation as a separate field, efficiency will remain an integral part of engineering, architecture and management practice. Therefore, one will continue to find much important information about efficiency in the various specialized references used by professionals in those fields.

The *ASHRAE Handbook*, published by the American Society of Heating, Refrigerating, and Air Conditioning Engineers, Inc., is the world's definitive reference for heating, ventilation, air conditioning and refrigeration. This reference covers the majority of the equipment and materials that determine energy consumption in the commercial and residential sectors, and in refrigeration applications. It is also the primary source of information related to the architectural aspects of building design. Energy efficiency is a major issue throughout the *Handbook*, but energy conservation measures are not presented explicitly.

The *ASHRAE Handbook* belongs in every reference collection that provides information about energy consumption in building construction and refrigeration. Its coverage of topics is thorough, and it is extraordinarily well organized. The depth of treatment varies among the topics, from thorough in many areas to superficial in a few. It is published in four volumes, each volume being revised annually on a four-year cycle through the work of many ASHRAE technical committees. This keeps it exceptionally current, in comparison to other major references. The two volumes dealing with systems and applications, respectively, are easy to read. The volume dealing with industrial refrigeration is detailed, but easy to read by specialists in that field. The volume dealing with the principles and theory of HVAC (called "Fundamentals") has come under criticism from practitioners for poor organization and increasing remoteness from practicality. Unfortunately, the indexing of the four volumes is weak.

For electric lighting, the prime reference source is the *Lighting Handbook*, published by the Illuminating Engineering Society of North America. This beautifully rendered book is a fundamental source of information about the technical details of lighting, including the efficiency characteristics of lighting equipment. It covers the field completely, it is well organized, and it is easy to read. Unfortunately, it is updated only at long intervals, and lighting technology is presently evolving rapidly at the level of detail covered by the book. Therefore, its information about specific equipment is likely to be dated (the current edition was published in 1999). Lighting design practitioners do not yet have a doctrine of lighting design that stresses efficiency. This defect is reflected in the *Lighting Handbook*. (The *Energy Efficiency Manual*, which also covers lighting extensively, threw down that gauntlet by discarding conventional lighting design methods and introducing a radically different approach that emphasizes efficiency.) Aside from failing to deal with efficiency, lighting design in the commercial and residential sectors is oriented heavily toward the use of light for cosmetic purposes, which increases energy consumption. Another factor that makes efficiency an orphan in lighting design is that responsibility for lighting design wanders between architects, electrical engineers, and anyone who wishes to call himself a lighting designer.

The *ASHRAE Handbook* and the *Lighting Handbook* are the primary design references used by the practitioners who determine most of the energy consumption in the commercial and residential sectors. It is noteworthy that architects have not produced a reference that deals adequately with energy efficiency in architecture. (The *ASHRAE Handbook* is a prime reference for the energy aspects of architectural design, but architects do not appear to use it.)

In the industrial sector, there are major references oriented toward mechanical engineers and electrical engineers. There are also major references oriented toward particular industries. *Marks' Standard Handbook of Mechanical Engineering*, published by McGraw-Hill, first appeared in 1916. Its audience is "heavy" mechanical engineering. Unfortunately, its age shows, and it has become a grab bag as mechanical engineering has diversified far beyond the content of the book. Efficiency is not a primary theme, and information related to efficiency tends to be intermingled with other issues. The latest edition is copyrighted 1987.

The *Standard Handbook for Electrical Engineers*, published by McGraw-Hill, is the original comprehensive reference for electrical engineers. It has little to say about energy conservation, in the sense that we defined it previously. Electrical engineering has diversified to the point that a single reference can no longer cover all aspects of it. There are now major references that deal with individual areas of electrical engineering. The design of electrical machinery is the aspect of electrical engineering that most affects energy conservation, and it is determined largely at the time the equipment is made.

Along with these general handbooks, there are major references dealing with particular industries. Perhaps the largest of these is the 25-volume *Kirk-Othmer Encyclopedia of Chemical Engineering*, published by John Wiley & Sons (1991-1997).

Professional and Trade Association Materials

A variety of professional and trade associations produce information that relates to energy conservation. Examples are the National Electrical Manufacturers Association (NEMA), the Association of Physical Plant Administrators of Universities and Colleges (APPA), the Electrical Power Research Institute (EPRI), and the Sheet Metal and Air Conditioning Contractors' National Association (SMACNA). In some cases, associations produce the primary information in their fields, which may have major relevance to energy conservation. Notable examples are the *ASHRAE Handbook* and the *Lighting Handbook*, which are both produced by engineering societies. Some trade associations produce consolidated information about the equipment and materials that their members manufacture. An outstanding example is the Air Conditioning and Refrigeration Institute, which publishes directories that give the efficiency characteristics of all packaged air conditioning equipment and heat pumps made in the United States.

Manufacturers' Catalogs and Training Materials

Most advances in the efficiency of equipment and materials are made by manufacturers. Therefore, manufacturers' literature is usually the first source of information about product efficiency that appears in a useful form. The best product information usually comes from one or a few manufacturers in each product area. These manufacturers assume the role of educator within their markets. Their goal is to influence the engineers, architects, contractors, and others who purchase or recommend products.

Manufacturers' best information may be contained in their catalogs, and also in pamphlets, design manuals, training aids, technical bulletins, and so forth. For example, Armstrong International and Spirax Sarco, two major manufacturers of equipment related to steam plant efficiency, offer excellent design guides for steam systems. One should be aware that manufacturers' literature tends to exaggerate efficiency performance. There is little outright lying in the catalogs of major manufacturers. However, it is a common practice to state equipment efficiency under ideal conditions, which may differ radically from typical operating conditions. For example, manufacturers of air conditioning equipment typically quote efficiency under full load, but the equipment operates at substantially lower average efficiency under typical load conditions. Another caution is that problems related to particular types of equipment and systems are swept under the rug. Adverse side effects are a general problem of energy conservation, but manufacturers avoid mentioning them, at least initially. For example, electronic motor speed controls are an important new product that can create serious electrical problems. Denial of the issue continues until a manufacturer can gain a competitive advantage by claiming the ability to tame the problem.

Magazines and Journals

There is little "scholarly literature" in energy conservation. Advances are made primarily by manufacturers, who keep their information proprietary until they are ready to release it. At that point, advances are announced in the manufacturers' own promotional literature, and in magazines of the respective trades and professions. The majority of magazines read by practitioners are free. These magazines earn their income by advertising, and the influence of advertisers is strong. The better magazines provide some degree of quality control. Others rigorously avoid subject matter that might offend anyone, especially advertisers. This is a serious defect because progress in energy conservation is now largely a matter of recognizing and solving problems.

A large fraction of magazine articles are written by manufacturers' marketing people, so the articles have the same credibility concerns as manufacturers' literature. Another fraction of articles are written by practitioners who are promoting their own successes. A related weakness of all magazines is a hankering for "success stories", to the point of censoring anything that sounds negative. This tendency makes magazine information misleading and incomplete. For example, the author once wrote an article for the journal of a major professional society to explain solar control films for windows. This technique has great appeal, but also serious deficiencies. Without informing the author, the editor deleted all reference to the deficiencies, turning the article into a puff piece. There are magazines and journals devoted specifically to energy conservation. Examples are Energy User News and Energy Manager. Magazines in this group tend to suffer from the "success story" fixation.

Valuable information about energy conservation appears in certain industry magazines. The trick is to know which magazines are the best in each field. For example, *Heating, Piping, and Air Conditioning* is perhaps the best in the field of commercial sector HVAC. It has a large editorial board of leading professional engineers, a quality control device that is virtually unique among the free magazines. More specialized information may be found in odd places. For example, *Store Equipment and Design*, which serves the food store industry, has an interesting monthly section that deals with energy conservation in food stores. The subject matter of magazines tends to appear in waves, as editors seek to exploit interest in current topics. The downside of this is that a good article about a topic that is not current may not appear in magazines for several years.

Conference Proceedings

Since energy conservation is an evolving field, a variety of conferences are held at which information is presented. Most of the conferences that produce good information occur annually. Leading examples of conferences devoted specifically to energy conservation are the World Energy Engineering Conference and the Industrial Energy Technology Conference. Many conferences oriented to specific industries produce information that relates to energy efficiency. A leading example is the semi-annual meetings of the American Society of Heating, Refrigerating, and Air Conditioning Engineers.

Conferences are organized by trade associations, universities, and companies that specialize in arranging conferences. Conference funding typically is provided by combinations of sponsors, typically associations and government agencies. Exhibits at conferences provide revenue from vendors, as well as attracting participants. A conference may have a dominant theme, or a variety of themes. Presentations are grouped into sessions. Session chairmen select speakers, who write the papers. The papers typically are written by practitioners, manufacturers' representatives and academics. The major conferences archive their papers, and some conferences collect all the presentations into bound volumes that are called "proceedings".

The average quality of conference information is good. Topics are treated in greater depth than in magazines. Speakers have considerable latitude to express their opinions, and there is no opportunity to escape from probing questions. The author of a paper faces personal exposure to a large number of peers, which is strong motivation to provide information that is relevant and credible. As a result, conference proceedings are a source of information about practical issues that may not be discussed candidly elsewhere. Conference proceedings tend to age well. This is because most energy conservation technology is not advancing rapidly, and because the same problems keep arising year after year. For this reason, you may find good information in conference proceedings that are several years old.

Newsletters

As the name implies, newsletters tend to be oriented toward news. Newsletters typically do not have advertisers, so they will tackle issues that magazines avoid. The most candid are produced by private individuals. Newsletters are also produced by universities, government agencies, associations and others.

Such newsletters tend to be more inhibited, or more selective, about controversial issues. The information in newsletters tends to be fresh because newsletters require considerably less preparation than other print sources. They are short, usually plain, and few people are involved. In fact, commercial newsletters (on all topics) may be produced by one person. By the same token, newsletters are narrower in scope than almost any other type of information source. Newsletters tend to be free or very expensive. Free newsletters must be subsidized in order to function. Commercial newsletters are expensive because of a lack of advertising revenue and a limited subscriber base. For example, the monthly *Energy Design Update* costs $ 337 per year, and *Environmental Building News* costs $ 179 for ten issues per year.

Government Sources

An enormous amount of information about energy conservation has been produced over the years by government agencies. The quality is variable. Most of the better information was produced by the U.S. federal government. State governments have sponsored much information, but typically it is derivative. The largest group of government information is consultant studies, which were produced in great quantity during the 1970s, and in reduced numbers thereafter. Many of these are available from the agencies who sponsored them, and from the National Technical Information Service (NTIS).

For the past two decades, the primary federal government source of information about conservation has been the U.S. Department of Energy (DOE). The DOE component primarily responsible for energy conservation is the Office of Energy Efficiency and Renewable Energy. (The name has changed over the years.) Information produced by the DOE reflects the organization of the administrative groups within this Office. The various groups produce information related to all the sectors of conservation that we described previously. For example, one group deals with specific industries and specific energy-using equipment, such as electric motors and air compressors. Another group deals with the residential and commercial issues that cover a much broader range of issues. A sub-group of the latter develops energy efficiency standards in response to federal law. And so forth. The Energy Information Administration (EIA) is part of the Department of Energy. The EIA is oriented primarily toward energy supplies, rather than conservation. Internet queries about energy conservation are diverted out of the website to a general search engine.

The U.S. Environmental Protection Agency (EPA) is primarily a regulatory agency responsible for enforcement of environment laws. Because fuel consumption is a major cause of air, water and ground pollution, EPA has become aggressively involved with energy conservation, increasingly taking the initiative for energy conservation away from the Department of Energy. As a result, EPA is becoming a source of information about conservation. An example is the equipment efficiency information resulting from the Energy Star program. The list of Internet sites at the end of this article leads to the important government sites.

The Best Ways to Search for Information

At the present time, there is no comprehensive index to information about energy conservation that is similar to *Index Medicus* for medicine or *Lexis/Nexis* for law. Therefore, you may have to invest significant effort to find the best sources. Exploit the aggregations of information that we have explored. Here are some tips for organizing the search. First, one should try to judge what kind of information source will satisfy one's requirement best. See Figure 2, Comparison of Energy Conservation Information Sources.

Books are Usually Best

There are good reasons why books have been the primary medium for information throughout human history. In contrast with other media, books provide information in finalized form. They present a broad view of a particular subject. A book usually organizes information well. It has a table of contents and an index, making the information within the book quickly accessible. The investment required to produce a book and the exposure that a book provides to the author motivate the best possible job in presenting the information. During production, a book is refined by passing through the hands of people with skills in various aspects of communicating knowledge. One can find books that cover almost any specific energy conservation topic exhaustively or one topic in context among related topics. Even if the topic is highly specific, it is usually best to learn about it in a book, which will provide the context that surrounds the topic.

There is a current viewpoint that the Internet has made the book obsolete as a source of information, or that the Internet makes books a secondary source. In fact, the Internet challenges none of the advantages of books, and it is an excellent way to find books. Books are usually easier to find than other sources of information, especially if one is new to the world of energy conservation. Therefore, we will first examine the ways to search for books about energy conservation.

Internet Book Dealers

Internet book dealers are good places to find books, provided that you know the title and/or the author. They are much less reliable when searching by subject matter. Even so, one may have better success searching “energy management” on a large book dealer's site than one will searching “energy management” on the Internet as a whole using a search engine. A serious weakness of Internet book catalogs (and other catalogs) is finding a particular topic within books that cover many topics. Such books may be the best source. However, book dealers typically provide an inadequate number of key words for searching comprehensive books. For example, a subject search for “cogeneration” in Amazon.com yielded mostly books in which that word appears in the title. Two-thirds of these books were out of print, and most of the remainder were special orders. Thus, one would miss many readily available books that address this popular topic.

Defective cataloging is another weakness of Internet booksellers. Books may be classified incorrectly, or they may be listed in categories that are not expected by people making searches. As of now, the major Internet dealers do not offer extensive descriptions of books. Limited information, such as tables of contents and reviews, may be provided. Be aware that "book reviews" on dealer sites are often written by claques. However, these may be the only reviews for an energy conservation book that one will be able to find.

Publishers' Websites and Catalogs

Once one finds a particular book that looks promising, one will get the most complete information about it from the publisher. Most catalogs list the publisher's name, but not all do. If the catalog does not list the publisher, it may list the ISBN. In any case, one can use an Internet search engine with the title of the book and/or the name of the author to find the publisher's website quickly. For example, by going to the Energy Institute Press website, **http://www.energybooks.com/**, you can view extensive information about the *Energy Efficiency Manual*, including the table of contents, the entire 40-page index (as a searchable PDF file), and extensive extracts from the book.

Book Catalogs Specializing in Energy Conservation

A few catalog booksellers specialize in books about energy conservation. In addition, catalog booksellers who specialize in related topics, such as construction or environmental protection, may carry a limited selection of books about conservation. Such catalogs are offered by commercial booksellers (e.g. Iris Communications), professional and trade associations (e.g. AEE and ASHRAE), advocacy groups and others. An advantage of these catalogs is that they are small enough to be examined in their entirety, so you can gain a good overview of what the catalog offers. Typically, a catalog offers brief descriptions of the books, making it easier for you to judge their relative merits. A disadvantage of specialty catalogs is that they offer only a fraction of the books that are available about their subject. The catalog may be produced by a publisher that sells only its own books. Or, the catalog may offer only the books of a limited number of publishers. Given their lack of completeness, why bother with specialty catalogs? Because not all good books about energy conservation are sold by the major booksellers.

General booksellers may not want to carry books that are considered to be of limited interest. Or, the publishers of the books may lack the sophistication to make their books available through the major booksellers.

Book Reviews

Judging the quality of a book by reading the publisher's description in a catalog is almost impossible. One may not even be able to judge the potential audience of the book. Independent book reviews are the time honored solution to this problem. Unfortunately, only a fraction of books about energy conservation ever receive serious reviews. Some review publications, such as *SciTech Book News*, announce the arrival of most books that are submitted to them. Unfortunately, such review publications tend to report contents in a bland manner that does not say much about the quality of a book. Also, most reviews do not tell you the relative merit of a book compared to other books on the same subject. There are databases of book reviews that you can use. These are generally subscription services. Searching the Internet yields spotty results for book reviews.

Literature Databases

Librarians are being inundated by vendors of literature databases. It is unlikely that any of these can do a good job of leading you to the best energy conservation books. It would take the entire population of a small country to generate a database that would allow one to find the best literature on any topic. And, as we said before, there are no databases that deal comprehensively with the subject of energy conservation alone. A database is no better than the people who do the indexing, and this function tends to be sloppy. Descriptions, if any, tend to be perfunctory. Even the mainstream directories, like *Books in Print*, are disappointing. For example, the *Energy Efficiency Manual* is listed under the single classification of "energy production", which is incorrect [this has since been corrected and the *Manual* has been given the subject "energy conservation" in the online version of *Books in Print*]. There are exceptions in some areas that are distantly related to energy conservation. For example, *Petroleum Abstracts* is excellent, but does not tell much about conservation.

Tips about Searching the Internet

The Internet has become a popular tool for finding books, articles, or almost any other form of information. One should know how to use it, and be alert to its weaknesses. The Internet is not yet a well organized source of information, and it is far from complete. It is a huge, growing dump into which information of differing quality and relevance is mixed indiscriminately. The Internet has no magical ability to capture information. In order for information to appear on the Internet, someone must create a website and put the information on the site. Much good information does not appear on the Internet. Even if information is located on the Internet, it may be difficult or virtually impossible to find. To find information, one must know the address (URL) of a particular website beforehand, or one must use search engines. Keep a record of URLs that have proven useful. Often, one good URL will link to others. Knowing the right URL for a topic may radically shorten the search time (see the list of websites at the end of this article).

The Internet is handy when searching for information about an unfamiliar subject. One can use a search engine, the tool that allows one to find the gems that are scattered throughout the Internet's information dump. A search engine is a computer program that matches search words that you specify against the vast amount of information on the Internet. There are several major search engines. They are offered free to Internet users. Each search engine operates somewhat differently, so each will yield different results in a search. However, the same search engine may be licensed under different names, so you may get identical results from search engines that you think are different.

How do search engines work? Each search engine continually sends out a search program called a "spider" that looks at the text of all websites. Using various criteria, the spiders decide whether or not a site is worth cataloging. If a site is deemed worthy, the spider will store fragments of text from various pages of the site, usually the home page. Some search engines use humans to refine the selection of sites and the information that is collected. This prior indexing is what allows search engines to find information with a speed that seems magical. Later, when an Internet user requests a search, the search engine scans its own catalog for a match and presents you with the corresponding web page fragments.

Search engine spiders look for two kinds of text entities on a website, the visible text and invisible "tags". If the text on the website does not match your search words, the search engine cannot lead you to that site. Poor search response for a particular website is often the fault of deficiencies in the site design. Tags, which are a part of the hidden background (HTML) code of the website, are critical for most search engines. Many websites have no tags, or the tags are poorly selected. In addition, the visible text of the website plays a major role in getting hits. There are other limitations that website developers may overlook, or they may be impractical to address. For example, important text that is displayed inside graphics boxes may not be seen by search engines. Also, search engines typically do not enter PDF files, which may contain the most relevant information on a site. To get a better understanding of how search engines look for Search Engine Watch: Tips About Internet Search Engines and Search Engines Submission <http://www.searchenginewatch.com/>.

There is a serious flaw in the present Internet system for tagging and searching websites. The present system limits the number of topics that can be listed in the invisible tags. And, if a topic is listed among many other topics, it gets lower search priority than the same topic would receive if it were the sole subject of a website. As a result, search engines often miss the best information sources, especially comprehensive books and conference proceedings. For this reason, one needs to know which are the best references in each subject area that you are searching. Understand that a search engine finds only content that is contained on a website. It rarely provides you with primary content, except for samples. For example, a web search may lead one to a particular book, but it will not give the content of the book to download. Information is a valuable commodity, and little is given away. Assuming that a reference to good information is present on the Internet, how well one finds it depends on two factors: which search words you use, and how cleverly the search engine matches your search words to the information that is posted on websites.

One fundamental problem of search engines is an excessive number of hits. One faces a needle-in-a-haystack problem, which is that good sources are lost among many irrelevant or low-quality hits. You can narrow your search by using more specific search words, but this may cause you to miss a valuable website because the site does not contain all your search words in proximity to each other. One can reduce the number of extraneous hits by using the "advanced" search features that search engines offer. These comprise various ways of combining and limiting search words. Each search engines provides easy directions for using these search features.

The following table is an example of the number of responses that one will get from typical searches. It illustrates how variations in search word inputs have a major effect on the results that one gets. (These searches were done with the search engine, northernlight.com.)

Search Words, Exactly as Written	Number of Hits
Industrial Energy Technology Conference	131,660
"Industrial Energy Technology Conference"	187
heat pumps	208,353
"heat pumps"	54,073
ground source "heat pumps"	11,054
geothermal "heat pumps" [Note: "geothermal" is a synonym for "ground source" in this context, but the hits obtained were substantially different.]	7,700
geothermal "heat pumps" principles	741
book reviews	3,522,494
"book reviews" AND "energy conservation"	615
Energy Efficiency Manual	47,673
"Energy Efficiency Manual"	50
book reviews AND "Energy Efficiency Manual"	9
"book reviews" AND "Energy Efficiency Manual"	none
"Energy Efficiency Manual" AND Wulfinghoff	8

Good search engines automatically search for variations of a word. So, if you search for "pump", the search engine will recognize "pump", "pumps", "pumped", and "pumping". However, do not expect search engines to recognize related concepts. For example, "energy efficiency" will turn up many books that do not appear in response to "energy conservation", and vice versa. Some search engines sort hits into labeled groups. This may radically improve your ability to find the most relevant hits. For example, in a search for information about heat pumps, one search engine divided the hits into fourteen groups. However, the grouping depends on the information contained in the individual websites, so it is no more reliable than the search itself.

In summary, the Internet is an excellent resource, and it is certainly the easiest. One does not have to get out of the chair to use it, and it provides information immediately. But, do not be seduced into using it exclusively. Only rarely will the Internet be your actual source of information about an energy conservation topic. It will fail to tell you about some of the best sources.

How to Get Information from Manufacturers

Manufacturers have a strong interest in being found, so there are many directories of manufacturers that are indexed by the types of equipment they offer. Some directories, such as the *Sweet's Catalog*, include the manufacturers' own catalogs for a fee. For the larger manufacturers, these are only small extracts. Expect to go directly to the manufacturer to find out what is offered. The question is, which manufacturer should one approach? Usually, only one or two manufacturers of each type of equipment offer good explanatory information. To identify the manufacturers with the best information, contact a leading practitioner in a related field. For example, to find out which manufacturers offer the best information about pump efficiency, contact a mechanical engineer who designs pumping systems. It may take a series of telephone calls to find such an individual. Or, if you are lucky, your information client may steer you to a good contact.

There have been attempts to produce directories of manufacturers who produce equipment and materials related to energy conservation. These directories are not complete, and they do not tell you who offers good information, but they may be a good place to start. Iris Communications offered one of the better directories of this type, but they appear to have abandoned it. Most major manufacturers now have websites, which a good search engine may find when you search for a particular product. The Internet may be the quickest way to identify a relevant manufacturer. However, the manufacturer's website is unlikely to have the information you need. For that, pick up the telephone and call the manufacturer. Whom should you contact when you call a manufacturer? Your best bet is usually the engineering department, or "applications engineering", which is a sales department that is composed of technical people. Odd as it may seem, corporate marketing departments are often surly and uncooperative, and many have little knowledge of their own products.

Use the PER Division Discussion List

The Petroleum and Energy Resources (PER) Division is the component of the Special Libraries Association (SLA) that is intended to serve librarians involved with energy conservation. The PER Division maintains an active "discussion list", which is a forum for posting e-mail messages. The list puts you in direct contact with a large fraction of the members of the PER Division. If you post a question on the list, you are likely to get a useful response within a day or two. This is a good way to start if you are stymied by an unusual question. At present, you do not have to be a member of the PER Division, or even a member of SLA, to use the discussion list. To join the list, go to the SLA PER Division website, http://www.sla.org/division/dper, click "Discussion List", and follow the simple instructions.

Provide Information, Avoid Misinformation

Progress in energy conservation has been crippled by misconceptions and misinformation. You are likely to encounter these in many of the requests for information that you receive. You can provide an important service to your client, and to society as a whole, by dealing professionally with these problems. You not only want to provide good information, but you want to avoid propagating misinformation.

What Does Your Client Really Want to Know?

As a provider of information about energy conservation, you often need to take a leadership position in guiding and formulating your client's request. This can be the most challenging aspect of your job. You will have to judge when to simply give your client what he requests, and when to probe to find what he really wants. As a rule, if your client comes to you with a very specific question, you are safe in proceeding directly to that information. An example is a question like, "Please find information about the relative efficiencies of single-stage and two-stage absorption chillers." In cases like this, just ask the client to explain his terms so that you search more intelligently. On the other hand, if your client comes to you with a question that sounds vague or general, he probably does not know what he wants. He may not even have a clear idea why he is seeking the information. He may be implicitly asking you to help define his request.

For example, during my career as an energy efficiency specialist, people have repeatedly approached me with the question, "Do you deal with solar energy?" In the end, it turned out that they did not have a specific interest in solar energy, but they were using the term "solar energy" as a surrogate for energy conservation in general. Usually, they really wanted to figure out how to make their facilities more efficient in a broad sense. When making the original request, they were fishing. If it is not clear what your client really wants to know, work with him to answer this short series of questions:

- **What is your motive for making the request?** Help the client to define his purpose. Is he trying to understand basic principles or to make improvements? For example, if your client is seeking information about geothermal heat pumps, does he merely want to know the principles of operation, or does he intend to install one? Information about the principles of geothermal heat pumps is widely available. However, you will have to search longer to discover that geothermal heat pumps are plagued by problems, and you will have to search even longer to learn how to minimize the problems. Or, was your client's interest in heat pumps recently stimulated by seeing a television program, reading an article, or attending a conference? In that case, the client's potential interest may be much broader than heat pumps.

- **What is your background?** Provide information that is appropriate for your client's ability to understand. Is your client an architect, a homeowner, an equipment manufacturer, or a college professor? Is your client a facility engineer trying to lower energy costs, or a student writing a term paper? Each of these clients is best served by a different selection of information sources.
- **Have you considered all your options?** The client typically focuses prematurely on a particular energy conservation activity. Make an effort to find the range of approaches that apply to your client's situation. For example, if your client asks about solar energy because he wants to reduce his heating costs, suggest that he consider all the other ways of reducing heating cost. Some of these methods may complement each other, and others are mutually exclusive.
- **How do your options compare?** Energy conservation activities differ widely in the benefit they provide, their cost, their reliability, and how easy they are to accomplish. For example, adding a geothermal heat pump to reduce the heating and cooling costs of a new house is expensive, and the system is likely to be unreliable. The same saving in heating and cooling cost may be possible by increasing the amount of insulation, which is simple, reliable, and inexpensive. Or, the same saving may result from upgrading to conventional equipment having higher efficiency.

Gadgets are Not Energy Conservation

A common misconception, even among energy professionals, is that energy conservation consists of installing the latest gadgets. Actually, effective energy conservation consists mostly of applying conventional equipment and materials reliably, and following efficient operating practices. Gadgets appear from time to time, but most of them prove to be impractical, riddled with problems, or pure snake oil. Some examples of questionable or useless gadgets include fluorescent fixture reflectors, fuel oil additives and electrical power surge suppressors. Some gadgets, such as turbulators to improve the efficiency of old boilers, may be legitimate, but they typically claim much larger benefits than the laws of physics allow. If you recognize that your client is treating a gadget as a surrogate for serious energy conservation, help your client to broaden his view to include the broad range of possibilities that may apply to his situation.

Deal with Novelty Appropriately

An issue that is related to fascination with gadgets is the appeal of novelty for its own sake, which is commonly expressed as "innovation". More generally, there is a fascination with energy conservation measures that are novel, expensive and complicated. You can help your client by pointing out that these three characteristics correlate to failure. In terms of achieving success, innovation is a serious liability. Things never work well at first, in any field of endeavor. The way to achieve successful conservation is to use methods that are well proven. There are a few areas where energy conservation is still awaiting major developments, but not many. Even in these areas, the prudent course is to wait until those developments have been well tested. Unfortunately, innovation has a powerful appeal, so your client may not be happy to hear you counsel against it. If your client requests information about emerging technology, find out whether he wants to understand the principles of the technology or to apply the technology. If your client's objective is to achieve reliable energy conservation, suggest that he skip the literature that is devoted to innovation.

Part of the appeal of innovation is the misconception that energy conservation is advancing at a blinding pace. This is not true. Most of the modern advances in equipment efficiency appeared during the 1970s. Most of the equipment available today settled into its present form during the 1980s. Equipment efficiency advanced radically during those two decades, and it is now largely stabilized. In most cases, your client does not have to worry about being left behind if he uses the best conventional technology. Most of the important development continuing at the present time is related to improving the reliability of energy-related equipment and to reducing adverse side effects. For example, electronic variable-speed drives for motors are one of the most important new advances in energy conservation. However, using them has the potential for creating serious problems in a facility's electrical system. There are various ways of dealing with these problems, which may be the most important information that you can provide about this technology.

In contrast with equipment evolution, which is largely complete and stable, design practice is still primitive with respect to efficiency. There is opportunity for radical improvement in all aspects of design practice, especially in architecture and lighting. Unfortunately, the information about efficient design in these areas is fragmentary and largely unsatisfactory. This is because the information is written largely by the people responsible for the problems.

Be Cautious with Advocacy Literature

Energy conservation is a technical subject with a strong emotional appeal for many non-technical people. This stems from the subject's novelty, its naturalistic aspects, a doomsday threat in the event that conservation fails, the intervention of government, and confusion of issues. As a result, information about energy conservation is an odd mix, ranging from serious professionalism to virtually religious beliefs. This diversity creates many niche markets for information, and advocacy groups compete to enter these markets. At the extremes of energy conservation are the professionals and the enthusiasts.

Roughly speaking, the professionals are those with technical training, who treat energy conservation as primarily an engineering issue. Enthusiasts commonly lack technical understanding, and they view energy conservation as a social and political activity. For some enthusiasts, energy conservation has become virtually a cult, especially when it is associated with environmental protection.

Carl Sagan once lamented that ten times more people are interested in astrology than in astronomy. A similar ratio applies to energy conservation. However, practitioners of astrology and astronomy isolate themselves from each other, and there is little confusion between the two interests. In contrast, professional interest and enthusiast interest overlap widely in energy conservation. Professionals generally cluster within professional organizations, out of public view. In contrast, enthusiasts tend to be represented by advocacy organizations, which strive for publicity. The information produced by advocacy organizations is heavily flavored for publicity. As a result, advocacy dominates the public discussion of conservation, and thereby the political action. Increasingly, the heads of government agencies involved with energy conservation and environmental protection are enthusiasts rather than professionals.

Most serious books about energy conservation are written by technical professionals, because they have enough knowledge to write good books. However, there is a substantial body of advocacy literature that is written by people with a weak grasp of the technical issues. Advocacy literature is characterized by blindness toward the practical issues that determine whether conservation will succeed. In advocacy literature, favored solutions to energy and environmental problems will work without question, if only obtuse and reactionary people would recognize their virtues. Scrutiny of actual energy savings and concern about cost and reliability are dismissed as resistance to progress. Enthusiasts favor certain esthetic characteristics. Favored methods tend to be unusual, small, local, and highly visual, with a rustic and residential flavor. Examples are solar systems, little wind generators, and straw bale houses. Things involving soil or mud are highly favored.

In these subject areas, much of the useful literature has been written by enthusiasts, who typically are specialists seeking to promote a particular type of energy conservation to which they have devoted a large part of their careers. The information tends to be technically correct, and it may be the best that is available. However, candor about critical shortcomings tends to be inversely proportional to the enthusiasm of the author. You will avoid giving misinformation if you recognize advocacy literature for what it is. At best, its technical validity is questionable. Be aware that advocacy sources commonly include energy conservation as part of a broader agenda, or they may use it as bait to attract support for an agenda that is not overtly stated.

Internet Sites

This is a selected list of Internet sites to get you started in searching for information about energy conservation. These comprise only a small fraction of the sites that may contain information useful to you. However, they are likely to satisfy your requirement, or to lead you to other sites that may do so.

Note that the prefix "http://" has been omitted from all website addresses for brevity and clarity.

Publishers of Books and Monographs Related to Energy Conservation

The following are some of the most important publishers of information about energy conservation, along with a sampling of general publishers of information in related fields. This list is far from complete, especially for books on individual topics.

Publisher	Website	Products
Energy Institute Press	www.energybooks.com	*Energy Efficiency Manual*, the primary reference and how-to guide for energy conservation.
Association of Energy Engineers	www.aeecenter.org	Many books about different aspects of energy conservation. Also some general construction topics.

Contd...

Publisher	Website	Products
American Society of Heating, Refrigerating, and Air Conditioning Engineers, Inc. (ASHRAE)	www.ashrae.org	*ASHRAE Handbook*, the primary design handbook for HVAC, refrigeration, and the energy characteristics of building architectural features. Also, books on related topics and primary standards.
Air Conditioning and Refrigeration Institute	www.ari.org	Directories of air conditioning and refrigeration products with efficiency ratings. Standards and guidelines. Training publications.
Illuminating Engineering Society of North America (IESNA)	www.iesna.org	*Lighting Handbook*, the primary handbook of lighting technology. Monographs on many aspects of lighting.
Electric Power Research to Institute (EPRI)	www.epri.com	Many monographs related the electric utility industry, including related energy conservation technologies.
International Energy Agency	www.oecdwash.org	Reports on world energy supplies and related environmental issues
McGraw-Hill Inc.	www.mcgrawhill.ca	Primary engineering references, including *Marks' Standard Hanbook for Mechanical Engineers*,
	Standard Hanbook for	*Electrical Engineers*, many energy-related books
Prentice Hall	www.phdirect.com	Many energy-related technical books
John Wiley and Sons	www.wiley.com	Many energy-related technical books
National Technical Information Service (NTIS)	www.ntis.gov	Publications produced by the U.S. government, including studies.
Sheet Metal & Air Conditioning Contractors' National Association	www.smacna.org	Primarily publications and standards related to duct design. Some books related to energy conservation.
Air Conditioning Contractors of America	www.acca.org	Primarily training publications related to air conditioning.

Books Catalogs Specializing in Energy Conservation and Related Topics

Very few catalogs specialize in energy conservation. A larger number of catalogs include energy conservation books, but usually not as a clearly identified category.

Organization	Website	Products
Association of Energy Engineers	www.aeecenter.com	Lists only books published by AEE.
Iris Communications, Inc.	www.oikos.com	Books related primarily to residential construction using efficient and renewable techniques.
American Society of Heating, Refrigerating, and Air Conditioning Engineers, Inc.	www.ashrae.org	*ASHRAE Handbook*, ASHRAE books on HVAC and building envelope topics, ASHRAE standards, conference proceedings, major books by other publishers
Illuminating Engineering Society of North America	www.iesna.orge	*IESNA Lighting Handbook*, IESNA technical monographs, books by other publishers on related topics
Construction Book Express	www.constructionbook.com	Many books about all aspects of construction. Does not identify the publishers.

Appliance and Equipment Efficiency Databases

The following Internet sites provide immediately useful information about the efficiency of a wide range of commonplace energy-using appliances and other products. [These sites were compiled by David Scott Smith, Executive Director of the White House Interagency Environmental Technology Office.]

American Council for an Energy Efficient Economy (ACEEE) Appliance Information

These are listings of the most efficient models of common types of appliances, compiled by the American Council for an Energy Efficient Economy.

Website	Appliance or Topic
hes.lbl.gov/hes/ACEEE/intro.html	General information about appliance efficiency standards, utility rebate programs, etc.
www.aceee.org/consumerguide/topfridge.htm	Refrigerators and Freezers
www.aceee.org/consumerguide/topdish.htm	Dishwashers
www.aceee.org/consumerguide/topwash.htm	Clothes Washers
www.aceee.org/consumerguide/toprac.htm	Room Air Conditioners
www.aceee.org/consumerguide/topcac.htm	Central Air Conditioning Systems
www.aceee.org/consumerguide/topwater.htm	Water Heaters
www.aceee.org/consumerguide/topgboil.htm	Gas Fired Heating Boilers
www.aceee.org/consumerguide/topoboil.htm	Oil Fired Heating Boilers
www.aceee.org/consumerguide/topfurn.htm	Forced Air Heating Furnaces
www.aceee.org/consumerguide/topcashp.htm	Air-source Heat Pumps
hes.lbl.gov/hes/ACEEE/post.html#info	Information about the ACEEE *Consumer Guide to Home Energy Savings*

California Energy Commission Database of High-Efficiency Appliances

This group of sites lists categories of high-efficiency appliances that EXCEED BY AT LEAST 15% current California and Federal appliance efficiency standards. In each listing, the models are listed in order of decreasing efficiency.

Website	Appliance or Topic
www.energy.ca.gov/efficiency/appliances/conditioners/CENA.HTM	Central air conditioners
www.energy.ca.gov/efficiency/appliances/roomair/ROOMAIR.HTM	Room air conditioners
www.energy.ca.gov/efficiency/appliances/heatpump/HPMP.HTM	Heat pumps
www.energy.ca.gov/efficiency/appliances/furnaces/FURNACE.HTM	Central gas furnaces
www.energy.ca.gov/efficiency/appliances/refrig-freezers/refrig-freezers.html	Refrigerators, refrigerator-freezers, freezers and wine cellars

California Energy Commission Appliance Efficiency Database

www.energy.ca.gov/efficiency/appliances

This is the main California Energy Commission database, of which the previous items are parts. The main database lists appliances that are certified to the CEC by their manufacturers to MEET (rather than exceed) the current energy efficiency standards. These appliance types are included:

A. Refrigerators, refrigerator-freezers, freezers and wine chillers (3,700+ models)
B. Plumbing fittings and fixtures (10,000 models)
C. Central air conditioners and heat pumps (13,500+ models)
D. Room A/C, package terminal A/C, package terminal heat pumps (3,700+ models)
E. Spot air conditioners and computer room air conditioners (600+ models)
F. Cooking and washing appliances (1,200+ models)
G. Central fan type furnaces (11,700+ models)
H. Gas space heaters (3,000+ models)
I. Boilers (7,200 models)
J. Pool heaters (200+ models)
K. Lighting and ventilation control devices (1,000+ models)
L. Fluorescent lamp ballasts (450+ models)
M. Water heaters (gas) (5,800+ models)
N. Water heaters (electric) (2,500+ models).

U.S. Department of Energy Lawrence Berkeley Laboratory Product Information

This group of sites comprises the "Home Energy Saver Library: Product Information". It includes even more categories of appliances and equipment than the previous California Energy Commission site. These products are labeled in the "Energy Star" program. This site includes a limited amount of narrative about the types of equipment, as well as describing the ratings.

Website	Type of Equipment
hes.lbl.gov/hes/prodinfo.html	Index page for Energy Star products and links to ACEEE product ratings
www.epa.gov/appdstar/hvac/prodfur.html	Furnaces
www.epa.gov/appdstar/hvac/prodcac.html	Central Air Conditioners
www.epa.gov/appdstar/hvac/prodghp.html	Geothermal Heat Pumps
www.epa.gov/appdstar/hvac/prodcac.html	Electric Air-Source Heat pumps
www.epa.gov/appdstar/hvac/prodgas.html	Gas-fired Heat Pumps
www.epa.gov/appdstar/hvac/prodthem.html	Programmable Thermostats
www.epa.gov/appdstar/hvac/prodblrs.html	Heating Boilers
www.epa.gov/appdstar/esoe/ database/clist.htm	Computers
www.epa.gov/appdstar/esoe/ database/mon.htm	Computer Monitors
www.epa.gov/appdstar/esoe/ database/plist.htm	Computer Printers
www.epa.gov/appdstar/esoe/database/ flist.htm	Fax Machines
www.epa.gov/appdstar/esoe/database/ xlist.htm	Copiers
www.energystar.gov/products/ clotheswashers/	Clothes Washers
www.energystar.gov/products/ dishwashers/	Dishwashers
www.energystar.gov/products/ refrigerators/	Refrigerators
www.energystar.gov/products/roomair/	Room Air Conditioners
www.epa.gov/appdstar/home_ electronics/products.html	Television Set and VCR's
www.epa.gov/appdstar/fixtures/ products.html	Residential Light Fixtures
www.energystar.gov/products/l windows/climate.htm	Windows (best choices by climate)
www.energystar.org/partners.asp	Manufacturers, Retailers, and Utilities who participate in the Energy Star program

Energy Conservation Advocacy and Government Energy Organizations

These organizations are devoted primarily to promoting energy conservation. Their information products change with time. The products listed here are their primary ones at present. Energy conservation is also promoted by many other organizations, not listed here, who are primarily concerned with other issues, such as environmental protection, property management, etc.

Organization	Website	Present Information Activities
Energy Institute	www.energy-institute.org	Publication of books, papers, and other media about primary technical and management issues of energy conservation.
Alliance to Save Energy	www.ase.org	Topical news, newsletters, position papers.
American Council for an Energy Efficient Economy	www.aceee.org	Appliance efficiency ratings, legislative alerts, press releases, position papers, LINKS to other sites, etc.
Energy Efficient Building Association Inc.	www.eeba.org	Conferences related to residential construction.
International Institute for Energy Conservation - U.S.	www.cerf.org/iiec	Newsletters, press releases, articles related to energy conservation in developing countries.
Export Council for Energy Efficiency	www.ecee.org	International market assessments, international events calendar, LINKS to other sites, etc.
U.S. Business Council for Sustainable Energy	www.bcse.org	Reports, quarterly newsletter.
The Results Center	www.crest.org/efficiency	Discussion groups about various topics, LINKS to other sites, etc.
U.S. Department of Energy Office of Energy Efficiency and	www.eren.doe.gov	"Access to 80,000 documents and more than 600 LINKS" about

Contd...

Organization	Website	Present Information Activities
Renewable Energy		conservation and renewable resources.
U.S. Department of Energy Office of Industrial Technologies	www.oit.doe.gov	Newsletter, press releases, conferences, conservation programs for specific industries, telephone inquiry line, etc.
U.S Department of Energy Energy Information Administration	www.eia.doe.gov	A wide variety of information energy supplies, but not much about conservation.
International Energy Agency Energy Technology Data Exchange	www.etde.org/edb/energy.html	"Contains the world's largest collection of energy literature, with more than 3.8 million abstracted and indexed records ... updated twice monthly"

Contributors

John Coates <jcoates@stu.ca>, Professor of Social Work, St. Thomas University, Fredericton, New Brunswick, E3B 5G3 Canada.

Terry Leahy Terry.Leahy@newcastle.edu.au, Senior Lecturer, School of Social Sciences, University of Newcastle, Callaghan NSW 2308, Australia.

Annie Muldoon <goinggreenottawa@yahoo.ca> is a graduate of Carleton University, Ottawa, Canada where she completed a Master of Social Work. 1125 Colonel By Drive, Ottawa, ON, K1S 5B6, Canada. She currently works as an environmental consultant for non-profit agencies, and as a counsellor.

Ryder W. Miller <dolphin1965@hotmail.com> is a freelance environmental and science reporter who has been published in Sierra Magazine, California Coast & Ocean, California Wild, and Hydrosphere.

Natalie Persadie <npersadie@ecolegal2001.com>, is a graduate of the University of the West Indies, St. Augustine, where she completed a BA in French and Portuguese, a Post-Graduate Diploma in International Relations, as well as a Master of Philosophy in International Relations with a thesis on Development and Environmental Protection in Trinidad and Tobago. Ms. Persadie is currently pursuing doctoral studies at the University of Birmingham, England.

Rajendra Ramlogan <rramlogan@ecolegal2001.com> is a graduate of the University of Cambridge, England, where he obtained his doctorate in international environmental law and a graduate of the New York University School of Law where he obtained his LL.M. in International Legal Studies. Dr. Ramlogan did his Bachelor of Laws degree and Bachelor of Arts degree in English Literature at the University of the West Indies where he is currently lecturing.

V. Sekeran <vsek01@yahoo.com>, Prof. of Civil Engineering., Alagappa Chettiar College of Engineering. and Technology, Karaikudi – 630 004 Tamil Nadu, India.

C. Balaji <envirojee@yahoo.com>, Post Graduate student in Environmental Engineering, Alagappa Chettiar College of Engineering and Technology, Karaikudi – 630 004, Tamil Nadu, India.

T. Bhagavathi Pushpa <bhagavathi_pushpa@yahoo.com>, Post Graduate student in Environmental Engineering, Alagappa Chettiar College of Engineering and Technology, Karaikudi–630 004, Tamil Nadu, India.

Alan Marshall <alannigelmarshall@yahoo.com>, Department of Environmental Humanities, School of Social Studies, Masaryk University, Gorkého 7, CZ-602 00 Brno, Czech Republic.

Brian J. English, Ph. D. <rainbowreefwarrior@yahoo.com>, International Public Policy and Management Program, University of Southern California, 2831 Ellendale Pl. #20, Los Angeles, CA 90007, USA.

Carl M. Hand <chand@valdosta.edu>, Dept. of Sociology, Anthropology, and Criminal Justice, Valdosta State University, Valdosta, GA 31698 USA.

Ginger Macheski <macheski@valdosta.edu>, Dept. of Sociology, Anthropology, and Criminal Justice, Valdosta State University, Valdosta, GA 31698 USA.

Nallathiga Ramakrishna <krishnram@hotmail.com>, School of Planning Centre for Environmental Planning and Technology, Ahmedabad 380 009 India. Nallathiga

Ramakrishna is now the Programme Manger at Y.B. Chavan Centre, Nariman Point, Mumbai (Bombay), 400021, India.

David Grierson <d.grierson@strath.ac.uk>, Ph.D. is an academic, and a former resident of Arcosanti. He is currently Director of Design Studies for the Building Design Engineering (BDE) course and Deputy Director of the Sustainability and Environmental Management Unit (SEMU) at the Department of Architecture, University of Strathclyde, Scotland. He has recently lectured and published on various aspects of the Arcosanti project in Amsterdam, Brno, Glasgow, Massachusetts, Manchester and Prague.

Rosemarie Russo rrusso1@earthlink.net, Dean of Biological and Health Sciences, Diablo Valley College, 321 Golf Club Drive, Pleasant Hills, CA 94520 and Adjunct Professor, Thomas Edison College and University of Maryland. USA.

Donald Wulfinghoff <dw@energybooks.com> is the president of Wulfinghoff Energy Services, Inc., Wheaton, Maryland, USA.